T0206087

Quality Engineering Techniques

Quality Engineering Techniques

An Innovative and Creative Process Model

Ramin Rostamkhani
Mahdi Karbasian

CRC Press
Taylor & Francis Group
Boca Raton London New York

CRC Press is an imprint of the
Taylor & Francis Group, an **informa** business

First edition published 2020
by CRC Press
6000 Broken Sound Parkway NW, Suite 300, Boca Raton, FL 33487-2742

and by CRC Press
2 Park Square, Milton Park, Abingdon, Oxon, OX14 4RN

© 2020 Taylor & Francis Group, LLC

CRC Press is an imprint of Taylor & Francis Group, LLC

Reasonable efforts have been made to publish reliable data and information, but the author and publisher cannot assume responsibility for the validity of all materials or the consequences of their use. The authors and publishers have attempted to trace the copyright holders of all material reproduced in this publication and apologize to copyright holders if permission to publish in this form has not been obtained. If any copyright material has not been acknowledged please write and let us know so we may rectify in any future reprint.

Except as permitted under U.S. Copyright Law, no part of this book may be reprinted, reproduced, transmitted, or utilized in any form by any electronic, mechanical, or other means, now known or hereafter invented, including photocopying, microfilming, and recording, or in any information storage or retrieval system, without written permission from the publishers.

For permission to photocopy or use material electronically from this work, access www.copyright.com or contact the Copyright Clearance Center, Inc. (CCC), 222 Rosewood Drive, Danvers, MA 01923, 978-750-8400. For works that are not available on CCC please contact mpkbookspermissions@tandf.co.uk

Trademark notice: Product or corporate names may be trademarks or registered trademarks, and are used only for identification and explanation without intent to infringe.

Library of Congress Cataloging-in-Publication Data

Names: Rostamkhani, Ramin, author. | Karbasian, Mahdi, author.
Title: Quality engineering techniques : an innovative and creative process
model / Ramin Rostamkhani, Mahdi Karbasian.
Description: First edition. | Boca Raton : CRC Press, 2020. | Includes
bibliographical references and index.
Identifiers: LCCN 2020004495 (print) | LCCN 2020004496 (ebook) | ISBN
9780367903817 (hardback) | ISBN 9781003042037 (ebook)
Subjects: LCSH: Quality control. | Industrial management. | Project
management.
Classification: LCC TS156 .R6725 2020 (print) | LCC TS156 (ebook) | DDC
658.4/013--dc23
LC record available at https://lccn.loc.gov/2020004495
LC ebook record available at https://lccn.loc.gov/2020004496

ISBN: 978-0-367-90381-7 (hbk)
ISBN: 978-0-367-50006-1 (pbk)
ISBN: 978-1-003-04203-7 (ebk)

Typeset in Times
by Deanta Global Publishing Services, Chennai, India

Contents

Preface

No one can deny the incredible pace of change and progress in today's industrial and complex world. A large amount of information is exchanged and little time is available to deal with it. Many industries and firms of small, medium, or large sizes have a profound desire to increase productivity and sustainability to gain a competitive position in the global market. One of the best tools for achieving this goal is to apply Quality Engineering Techniques (QET). Quality Engineering Techniques can be established through process-oriented models applicable to all traditional processes employed in companies or firms. The authors of this book, having had more than 20 years of intensive work in Quality Management Systems (QMS) and Integrated Management Systems (IMS), have tried to share their applied knowledge and experience on the process-oriented model of quality engineering techniques with experts and managers working at different companies or firms, particularly those in industrial factories. The essential role of employing statistical techniques as the main tool of quality engineering techniques in the growth and development of industries is a well-known fact to the specialists in the field. In applying statistical techniques, the most referenced books in the field are those by Professor Douglas Montgomery from the Arizona State University (ASU) in the United States. In writing this current book, however, we have been inspired by well-known and accomplished professors in industrial engineering in Iran—Professor Rasoul Noorossana in particular—from the Iranian University of Science and Technology. Indeed, the invaluable books and articles he has published could improve knowledge of Quality Engineering Techniques for the first time in Iran. Furthermore, we have benefited from the articles and books published by Professor Arash Shahin from the University of Isfahan; in fact, the important progress and proliferation of quality engineering techniques in Iran have given us enough motivation to write this book. To the above authorities, we have to add the professors and researchers at the Malek Ashtar University of Technology whose great contribution to developing different levels of design and implementation of the process-oriented model are to be appreciated. Also, many experts and managers within this academic organization have cooperated with us in implementing the model presented in this book through applying it to defense sectors. We are indebted to them for their sincere efforts. Special thanks are due to internal and external participants in filling out the questionnaires. We are grateful to our colleagues who played active roles in developing different stages of the model. This model has been applied to a selected industrial factory. The obtained results, however, can be generalized not only to other industries but also to general service sectors. This book introduces, for the first time, an integrated and applicable model for quality engineering techniques and numerical applications for its implementation. It is to be noted that the design of the proposed model is introduced as the main core of the research project while its implementation is indicated in numerical application parts. So, the main thrust of our research is to provide answers to the important and essential questions listed below.

- How can we define the main specifications of a productive and sustainable model of QET in industrial organizations?
- How can different levels of the proposed model be configured?
- Which units or individuals are responsible for implementing the intended model of QET?
- How can the impacts of implementing the proposed model in question be assessed?
- Can this model create or increase added values for industrial organizations/firms?

The proposed model can prove useful to experts and managers who desire to achieve optimum productivity and sustainability through applying quality engineering techniques, whether statistical or non-statistical. The model presented can manage the application of quality engineering techniques in an integrated format for organizational processes. The most creative feature of the presented work is the idea of introducing a process map within an organization, besides exploiting several quality engineering techniques including statistical and non-statistical tools for different levels of the organization. The most innovative dimension of this book is executing the proposed model in an effective format for the defense sectors of Iran that can be generalized to other non-military sectors in each country as well as creating or augmenting added values for the manufacture of industrial products. We firmly believe that our model can be further improved by accommodating constructive experts' views from those authorities working in the manufacturing and general services sectors of organizations/firms. Please, do not hesitate to contact us (see below) and share your highly appreciated comments in regards to the content of the book.

Ramin Rostamkhani and Mahdi Karbasian
March 2019

About the Authors

Ramin Rostamkhani has an M.A. in Industrial Engineering. He earned his master's degree from the Malek Ashtar University of Technology in Tehran, Iran. He has worked in the Defense Industries Organization (DIO) in Iran. He has more than 20 years of experience in QET. Moreover, he has a lot of experience in QMS and IMS. He has written at least four articles in Scopus and ISI journals. He has extensive expertise and experience in the following areas:

- Reliability
- Productivity
- Sustainability
- Quality control
- Applied statistics
- Quality assurance
- Quality engineering
- Statistical and non-statistical techniques

Researchgate	https://www.researchgate.net/profile/Ramin_Rostamkhani/publications
LinkedIn	https://www.linkedin.com/in/ramin-rostamkhani-80446910a/

Mahdi Karbasian has a Ph.D. in Industrial Engineering. He earned a Ph.D. from the Tarbiat Modares University of Tehran, Iran. He is an Associate Professor in the Faculty of Industrial Engineering at the Malek Ashtar University of Technology in Esfahan, Iran. He is a manager of many quality projects in the Malek Ashtar University of Technology. He has written many articles in Scopus and ISI journals. He has extensive expertise and experience in the following areas:

- Process safety
- Failure analysis
- Applied statistics
- Reliability analysis
- Quality engineering

Researchgate	https://www.researchgate.net/profile/Mahdi_karbasian/publications
LinkedIn	https://www.linkedin.com/in/mahdi-karbassian-686ab235/

1 A Review of the Basic Concepts

1.1 INTRODUCTION

In the present century, quality engineering techniques have turned into applicable and effective tools for attaining advanced design and manufacturing technology as well as mass-production processes. The rationale behind these techniques from the very beginning was to help mass-production lines. It was only later that such methods were developed into useful instruments for other activities in the organization (both for pre-production activities such as product design, and for subsequent activities like after-sales services. Appropriate techniques are developed depending on the type of organization. The evolution path for quality engineering techniques (QET) (passing from low to high quality) has occurred synchronously with development in manufacturing lines. Various techniques have been developed and applied at each stage of the formation of a product (depending on the organization). However, it is important to note that most of the techniques are based on systematic processes, i.e., fewer inputs for converting qualitative outputs into quantitative ones. The result is that these techniques can establish a safe platform for decision making. That is to say, although these techniques can be applied individually, the logical nature is that when they are applied one after another, they act as reinforcements and exhibit double effects. The importance of the functional role of statistical techniques as a main core of QET for robust analysis of the data related to the indices of the strategic issues of quality management systems cannot be easily overlooked. The World Organization for Standardization, through one of its subcommittees, has shed light on identifying statistical techniques. This informative manual appears in two editions, in 1999 and 2003, where it is officially designated as ISO10017 which applies to all standards in the ISO9000 family, especially to those in ISO9001. This standard is a very useful tool in the identification of statistical techniques in the deployment, maintenance, improvement, and development of quality management systems. Statistical techniques as a mathematical tool in quality engineering play a crucial role in measuring, describing, analyzing, interpreting, and modeling system changes even with limited data. Statistical analyses in data can help us understand the extent and causes of changes. Hence, statistical techniques can prove beneficial in exploiting available data to help with decision making and to continuously improve the quality of products and processes, eventually improving customer satisfaction which is the most important goal of the organization. These techniques can be applied to an extensive range of activities such as market research, design, development, production, verification, and servicing.

1.2   THE HISTORY OF RESEARCH IN QET

Statistical control charts were employed by Shewhart for the first time in 1920. However, the relevant sciences flourished in the years between the two world wars. Later years witnessed the use of statistical quality control techniques in manufacturing military equipment. In the years after the war, i.e., in the 1950s and 1960s, the experiments, designs, and analysis techniques were also used. This time not only the military industries, but also the major automobile companies and their part makers turned to these sciences. Major companies like General Electric, General Motors, and Motorola provided a new era for testing these statistical techniques and even resorted to other engineering techniques that did not require strong statistical bases (non-statistical techniques) (Rezaei, 2001). In the field of statistical and non-statistical techniques, a great number of studies have been carried out whose implementation records can be seen in numerous references. It seems that a perfect application grounds for designing and implementing of QET can be realized in the defense sector where widespread and consistent processes are observable at three levels of processes: namely, main, leadership, and support procedures. One particular study discusses the advantage of applying statistical and non-statistical techniques related to quality engineering in the form of an integrated model for creating productivity and sustainability in the main domains of industrial engineering associated with manufacturing factories. (Karimi Gavareshki et al., 2018). The popular statistical techniques based on ISO10017 and non-statistical techniques are presented in Figure 1.1.

Figure 1.2 presents another categorization for QET.

1.2.1   THE HISTORY OF RESEARCH IN STATISTICAL TECHNIQUES

The family of statistical techniques that can help an organization achieve its objectives is as follows: (ISO10017:2003)

   A. Descriptive statistics
   B. Design and analysis of experiments
   C. Statistical hypothesis tests
   D. Process capability analysis
   E. Regression analysis
   F. Reliability analysis
   G. Sampling
   H. Simulation
   I. Statistical process control charts
   J. Statistical tolerances
   K. Time series analysis

1.2.1.1   Descriptive Statistics

Descriptive statistics refers to the methods employed to summarize quantitative data in such a way as to define the characteristics of data distribution. The characteristics of data mostly taken into consideration are the central value of data (e.g., averages);

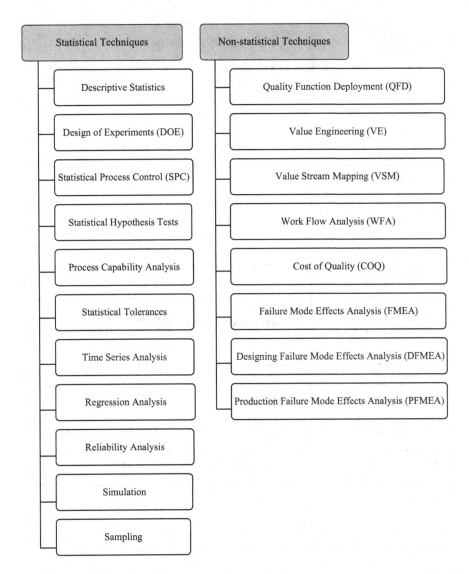

FIGURE 1.1   QET; statistical and non-statistical techniques.

and the dispersion of data (e.g., domains or standard deviations). Another feature of interest is the shape of the distribution of data (e.g., symmetries). The information obtained from descriptive statistics can often be easily and effectively influenced by resorting to various types of graphical methods including histogram charts, Pareto graphs, dispersion charts, causation charts or trend graphs. These graphical methods are useful since they are capable of discovering unusual aspects in the data that are vague in quantitative analyses. These methods are widely used in data analysis when the researcher decides to discover or verify the relationship among variables and intend to estimate the parameters used to describe these relationships.

QUALITY ENGINEERING TECHNIQUES		
LEAN TOOLS	**PLANNING TOOLS**	**ANALYTHICAL TOOLS**
Value Stream Mapping	Matrix Diagrams	SPC
5S	Responsibility Matrix	Pareto Diagrams
Poka Yoke		Cause & Effect Diagrams
Take Time		Histograms
Kaizen		Scatter Diagrams
Work Flow Analysis		Process Flow Charting
CREATIVITY AND THINKING TOOLS	**MONITORING TOOLS**	**DESIGN & DEVELOPMENT TOOLS**
Brain Storming	Statistical Process Control	FMEA
Nominal Group	Trend Charts	DFMEA
Practical Technique	Histograms	Design of Experiments
Affinity Diagram	Gantt Charts / Time Lines	Voice of the Customer
5 Whys		
Force Field		

FIGURE 1.2   QET.

1.2.1.1.1   Applications

In general, descriptive statistics are used to summarize and describe data attributes. This method is normally the first step in analyzing quantitative data. Therefore, as a first step and introduction to each analysis, these statistical methods are utilized. Examples of such applications are as follows:

- Summarizing the key indices of product features (e.g., average and standard deviations).
- Describing the function of some process parameters (e.g., temperature)

- Describing the delivery and response times (e.g., services)
- Summarizing the customer evaluation date (e.g., satisfaction or dissatisfaction)
- Displaying the measuring data (e.g., equipment calibration data)
- Displaying the distribution of features related to a process through a histogram
- Viewing the performance results of a given product over some time through a trend graph
- Evaluating the relationship between independent variables (e.g., temperature) and the output of that process as a dependent variable

1.2.1.1.2 Advantages
Using descriptive statistics is a convenient and simple way to summarize and describe data. It is also a good choice of procedure for providing information, especially by supplying graphical means for data and transferring information. Furthermore, this method is helpful in analyzing and interpreting data, which proves useful in making decisions.

1.2.1.1.3 Disadvantages
Descriptive statistics provides characteristics of sample data (for instance, means and standard deviations). However, these tools are contingent upon limitations such as sample size, and sampling method. These quantitative tools are considered valid when considered in relation to statistical assumptions.

1.2.1.2 Design and Analysis of Experiments
The design and analysis of experiments refer to all studies that are planned and carried out based on statistical calculations related to the results at a specified level. This technique involves making changes to the system under investigation and accordingly evaluating the effect of these changes on the system. Verifying some features of a system or examining the effect of one or more factors on these features can be defined as another goal of this technique. The arrangement and the tests that are conducted for this technique are extremely dependent on the purpose and the conditions of testing. There are various supplementary tools for analyzing data from variance analysis perspective such as checking the probability of points with having graphical natures.

1.2.1.2.1 Applications
Descriptive statistics can be used to evaluate the assessments of a product, process, or a system for verifying a specified standard or evaluating the comparisons made of several systems at a certain level. Confirmation of the effect of medical treatments and agricultural products, and evaluation of various types of methods in industrial productions are among the practical applications of this technique. The most practical aspect of this technique is its ability to examine complex systems whose outputs may be affected by multiple potential factors. As such, the purpose of the design of experiments under this condition is to optimize a feature or reduce its variability. In this case, descriptive statistics is used to analyze the factors that have the greatest impacts on the characteristics of the system. The results may be used to facilitate

the design and development of a product or process to control or improve an existing system. Examples of this may be to control or improve the average or reduce the variability in certain process characteristics such as process efficiency, product strength, or durability in factory products manufactured, for instance, by electronics, automotive, or chemical industries.

1.2.1.2.2   Advantages

One of the most striking advantages of designing and analyzing experiments is the creation of high-efficiency, economical procedure to examine the effects of several factors in a process, compared to the study of these factors. Also, the ability of this technique to identify the interactions between certain factors can lead to a deeper understanding of the process. Using the correct method of applying this technique, the risk of error in finding a random relationship between two or more variables is considerably reduced.

1.2.1.2.3   Disadvantages

There are some levels of variability inherent in all systems, which in some cases can prevent the attainment of accurate conclusions. While there may be misleading effects of some unknown factors, as well as the interactive effects of various factors in a system, choosing the right sample size and including other considerations might reduce the risk of errors in the final conclusions of the technique making it an acceptable outcome, although they cannot be totally eliminated. And in such cases, extending the generalization of the technique should always be limited to the internal workings of the selected scope.

1.2.1.3   Statistical Hypothesis Tests

This technique is a statistical method with a predetermined level of risk determining whether a set of data (typically from a sample) is compatible with a particular hypothesis or not. The hypothesis in question may apply to a specific distribution or statistical model and asks whether data is in the multiplicity of the parameters related to a particular distribution (for example, a mean value). As such, the statistical hypothesis test involves evaluating evidence to decide whether the hypothesis formed for a statistical model or parameter is to be accepted or rejected.

1.2.1.3.1   Applications

A hypothesis test generally decides whether a hypothesis on a parameter of a particular population (at a certain level estimated from a sample) is valid or not.

This testing technique is used to address the following questions/statements (given as examples):

- Does the average (or standard deviation) of a community show a certain amount?
- Does the meaning of two or more populations differ when comparisons are made?
- Is the proportion of defective items in a community greater than a certain amount?

- Does the distribution of a population show a normal curve?
- Are the samples taken from a community random?
- Testing the difference between the proportions of defective items against two process outputs
- Determining the sample size required to accept or reject a hypothesis at a specified level of certainty

1.2.1.3.2 Advantages

This technique claims to assess some parameter of a community with a certain level of certainty. Hence, it can be useful in making decisions contingent upon that parameter. As well, the method can provide useful information on the nature of the distribution of a community together with the characteristics of the sample data.

1.2.1.3.3 Disadvantages

Generally, to ensure the accuracy of the results related to the statistical assumptions, the samples should be considered independently and randomly. Further, although the level of assurance related to the results is obtained according to the sample, the assurance of independent and random sampling is not possible.

1.2.1.4 Process Capability Analysis

This technique can assess changes and distribution of a process to estimate the ability of outputs that are in conformity with the range of permissible changes. If the data are measurable variables (from the product or process), the measurements are to be determined through standard deviations of the process distribution, provided that they are under the control of the process's intrinsic variability. If the process data follows a normal distribution, this will be 99.73% of the statistical population. Generally, process capability is based on indices designated as C_P, C_{PK}, P_P, and P_{PK}. The first and the second indices with differences in process centering are used to measure the actual process variability, while the third and the fourth indices with the same difference in process centeredness are used to measure the overall process variability. Other indices for the process capability are designed to calculate the long-term or short-term variability as well as the variability around the intended target amount of the process. Moreover, if the process data includes parameters such as the percentage of non-conformities or the number of non-conforming items, the process capability is to be expressed in terms of average non-conformance rate or the average ratio of non-conforming items.

1.2.1.4.1 Applications

The technique can be used to create quality engineering specifications for manufacturing products that are compatible with the tolerance in the assembled parts. Also, it is used to achieve high quality as well as optimum cumulative reliability in complex systems. Hence, manufacturers of cars, aircraft, electrical and electronic equipment, food, medicine, and medical supplies make use of this technique as an important tool for evaluating the process of their production.

1.2.1.4.2   Advantages

In general, this technique evaluates the inherent variability of a process and estimates the percentage of the expected non-conforming items. Therefore, this assessment enables the organization to estimate the costs of non-conformity and to orientate decisions related to process improvement. As a result, the organization is informed in choosing the processes and equipment that would produce an acceptable product. Besides, it enables the manufacturer to apply minimal direct inspection of the purchased products and materials.

1.2.1.4.3   Disadvantages

Although this technique has great potential for evaluating the power of a process, the concept of process capability relies on the following assumptions:

- The process should be under statistical control.
- The statistical population under study should be normal
- Processes that have systemic causes (e.g., instrumental depreciation) should be avoided. Besides any of the above-mentioned problems, the effectiveness of this technique has its own drawbacks which necessitate careful attention to the use of compensatory methods.

1.2.1.5   Regression Analysis

The regression analysis determines the relationship between the behavior of a characteristic cause (response variables), and a potential cause (descriptive variables). Hence, the technique aims at understanding the potential causes of change in the response while determining the contribution of each of these factors accomplished through establishing a statistical relationship between the changes in the response variables and changes in the descriptive variables. The analysis can indicate the most effective method of minimizing the difference between the real and the ideal answers.

1.2.1.5.1   Applications

Regression analysis has the following applications:

- Examination of the assumptions related to the effect of independent descriptive variables on dependent variables (response) and predicting the value of the dependent variables (response) for the values of these independent descriptive variables. (Identifying the most important factors in processes and assessing their contributions to the variability of the desired features as well as forecasting the outputs of a test or study related to the past, present, and future conditions of a specified production.)
- Estimation of the direction and degree of the relationship between a dependent variable (response) and an independent descriptive variable. Of course, this does not imply the existence of a cause-and-effect relationship. (For example, determining the effect of changing a factor like temperature on the output of the process while the other factors remain constant.)

- Modeling different characteristics of a process. (For example, efficiency, output, performance, cycle time, probability of failure of the test, and various stoppages in the process.)
- Verifying the replacement of a measurement method. (Such as replacing a time-consuming method with a faster and more accurate one.)
- Non-linear applications. (For example, obtaining a production formula for a product as a function of time and volume of demand or obtaining a formula for a chemical interaction as a function of time, temperature, and pressure.)

1.2.1.5.2 Advantages

Regression analysis can provide the relationship between various factors and the desired response, and thereby help in the making of a decision related to the process under study, and can ultimately improve the process. The main capability of this technique is to accurately describe the patterns of response data, to compare differences and explain related sets of data, and to provide the acceptable estimate of the impact of independent variables on a dependent variable (the response). This type of information aids the realm of controlling or improving the outputs of a process. The regression technique can also estimate the response rate in a satisfactory manner as well as the source of the effects of factors that are either not measured or eliminated in the analysis. In general, the analytical capability of the technique, especially in predicting the effect of independent variables on a given response, can prove useful, especially for processes requiring time and cost.

1.2.1.5.3 Disadvantages

The use of regression analysis for modeling linear, exponential, multivariate, and other processes, in the absence of sufficient skill and experience in those working with the model, can lead to measurement errors and other sources of changes that can make the structured model too complicated. In some cases, as well, for creating a model, the accuracy of the available data may not be taken into consideration while checking the accuracy of such data is essential. As such, adding or removing this type of data from the analysis causes an incorrect estimation of the parameters related to the model, consequently affecting the response. Another important point is the existence of additional independent descriptive variables which can also prevent the discovery of the real effect of other independent descriptive variables on the dependent variable (response), whose elimination may seriously damage the validity of the model's results.

1.2.1.6 Reliability Analysis

Reliability analysis makes use of analytical and engineering methods for evaluating, predicting, and ensuring the correct operation of a product or system under study over time. The techniques used in reliability analyses often require the use of statistical methods to resolve uncertainties, random attributes, or probabilities of failure, etc. In this kind of analysis, parameters such as the time to failure or the

time between failures are dealt with. The technique includes other techniques like
analyzing the malfunctions and their effects which focus on the physical nature and
causes of failures.

1.2.1.6.1   Applications
Reliability analysis applies to:

- Validating key indices related to the reliability and predictability of the
  performance of various components and systems (e.g., time to failure
  or time between failures for a certain number of test units in a limited
  time)
- Providing statistical data for design parameters for predicting the cost
  of the product life cycle, as a consequence of which a new product is
  introduced
- Identifying critical components or parts of a high-risk process accurately
  to discover the causes of the product failure or weaknesses in the implementation of processes eventually providing the necessary background for
  corrective and preventive measures
- Supplying guidance for making decisions on manufacturing or buying general products
- Determining the major characteristics of product degradation to improve
  the product design or scheduling appropriately for the maintenance and
  repair

1.2.1.6.2   Advantages
- Creating the ability to correctly predict the desired performance of a prod-
  uct or process
- Achieving a plan on various parameters for designing a product
- Establishing objective criteria for the rejection or acceptance of conducting
  conformity tests for a product or system
- Planning appropriately for optimal timing and preventive replacement
- Realizing an accurate estimate of the cost-effectiveness of a new product
  design or system design

1.2.1.6.3   Disadvantages
One of the basic assumptions of this technique is that the performance of the
product or system under study should be satisfactorily followed by a specific
statistical distribution. Due to a lack of attention to the precise determination
of this statistical distribution, the accuracy of the estimates will be challenged
when the accuracy of the product or system performance is concerned. Also, the
issue becomes much more complicated when several failures affecting the product or system are involved. Also, if the number of the observed failures in a test
is suspiciously low, this might negatively affect the accuracy of reliability estimates. The testing carried out under this condition would put the results of this
technique in doubt and the uncertainty about predictions make by the method
would increase.

1.2.1.7   Sampling

Sampling is defined as a systematic statistical method for obtaining information about some of the characteristics of a community, by studying a part which represents the whole. Different methods are employed for sampling:

- Random sampling
- Systematic sampling
- Sampling successively

The way a method is chosen is determined by the purpose and conditions of the research.

1.2.1.7.1   Applications

Sampling can be divided into two general categories:

- Sampling for acceptance (inspection)
- Sampling for review

In sampling to accept or reject a group of items based on the results of the sample(s), the industry application is to provide certain levels of assurance and information about whether the inputs of a product or process can meet the necessary requirements or not. In a sample for review, a numerical or analytic study is used to estimate the values of one or more characteristics of a community. At this stage, we often deal with surveys in which information is collected on a particular topic (e.g., measuring customer satisfaction). Also, the method is used to determine the number of samples needed to measure one or more characteristics of the statistical community. Other aspects of the application of sampling through this method are

- Estimation of the proportion of a community that might purchase a particular product
- Estimation of the percentage of items that provide a measure for certain criteria
- Control of a production process involving operators, machines, and products for monitoring variability and determining corrective and preventive measures

1.2.1.7.2   Advantages

An appropriate sampling plan, compared to a census of the entire community or a 100% inspection, can certainly save time, cost, and labor. Additionally, sampling is the only way to obtain the right information when the product inspection involves destructive tests.

1.2.1.7.3   Disadvantages

In designing a sampling process, the following should be considered:

- How to select the sample size
- Sampling time
- Sampling method
- Basis for sub-grouping

However, failure to pay attention to any of the above factors, which are mostly disregarded, gives rise to error rates.

1.2.1.8   Simulation

Simulation is an execution method through which a system (theoretical or empirical) is mathematically presented in the form of a computer program so that it can solve a problem. If the method of presenting includes concepts of probability theory, especially random variables, the designation *Monte Carlo* method simulation is used.

1.2.1.8.1   Applications

In the field of theoretical sciences, this technique is used when no comprehensive theory of problem solving is known, or if one is known, it cannot be applied to strengthen this technique (space programs or advanced missile defense projects can be cited as an example). In the field of empirical sciences, the technique is used when a system or a process can be properly described with the help of a computer program. The following are some of the more specific uses of the technique:

- Modeling oscillation in advanced mechanical parts
- Modeling oscillation of component profiles in complex assemblies
- Determining optimal timing for preventive maintenance
- Cost analysis and other analyses in design and production areas for resource allocation

1.2.1.8.2   Advantages

In the field of theoretical sciences, simulation (especially the *Monte Carlo* method) provides an appropriate tool for solving problems, especially in cases where direct and straightforward computations might be very difficult to accomplish. In the field of empirical science, simulations are used for a variety of tests found to be experimentally impossible or very costly to conduct. Hence, simulation has the advantage of offering the best solution at the shortest time and lowest costs.

1.2.1.8.3   Disadvantages

Note that in the field of theoretical sciences, evidence based on conceptual reasoning is more useful than simulation techniques since the technique often does not show the reasons for the outcome result. In the field of empirical sciences, there might exist some limitations where the simulated model does not fit. For this reason, the method is not to be used as a suitable substitute for reviews and evaluations.

1.2.1.9   Statistical Process Control Charts

Process control charts—a graphic representation of the data—are drawn from the samples gathered periodically from a process and displayed on the graph in the time order they were collected in. The control limits in these charts show the intrinsic variability of a process in a stable state as the role of control charts is to help to assess the stability of a process carried out by examining punctuated data relative to the control limits. In the case of variable data, a control diagram is used to monitor the changes of the process centre and a separate control diagram is used to monitor the process fluctuations. For descriptive data, control charts typically represent the number or ratio of non-conforming items or the number of observed non-conformities in the samples taken from the process. The general pattern of these charts is *Shewhart* model variables. There are other examples of control charts that have specific features (such as moving average charts).

1.2.1.9.1   Applications

These charts are used to specify changes in a process where the recorded data is compared with the control limits. In the simplest possible way, a point outside of the control limits indicates a change in the process which might be attributed to some specific causes. These causes need to be analyzed and determined for the observation task outside of the control limits. Many organizations such as automotive, electronic, and defense industries often use this technique to meet two purposes:

- To prove the sustainability of a production process and its continuous sustainability
- To determine the risk and the scope of corrective actions

This useful technique is exploited in the machining industry to reduce unnecessary interferences in a process. Another aspect of this technique is the control of such typical features as average response time, error rate, and the frequency of complaints for measuring, complicating, and improving the performance in the service industry.

1.2.1.9.2   Advantages

Besides showing the data, the process control charts have uses in helping to find the right answer for the reason behind process fluctuations. The crucial point is to distinguish between randomized (inherent) fluctuations and fluctuations in certain cases. The following can be mentioned as important benefits of these graphs:

- Process control
- Process capability analysis
- Measurement system analysis
- Cause and effect analysis
- Continuous improvement

1.2.1.9.3   Disadvantages

The most important point in the useful application of these charts is the selection of logical sub-groups that form the basis for the effective use of the charts. The interpretation of these charts in identifying the sources of a process variability is also very important an in some cases it is overlooked. Thus, the outcome results might be misleading. Further, there are some short-term processes that have scant data for determining appropriate control limits. Another setback is the existence of alpha and beta errors that never approximate zero.

1.2.1.10   Statistical Tolerances

Statistical tolerance is a method of execution using certain statistical principles as a basis, which is applied to determine tolerances form a two-sided viewpoint.

1.2.1.10.1   Applications

In cases where multiple individual parts or members are assembled in a single unit, the final value, using this technique, occurs only when the dimensions of all the individual sectors are located at the bottom or above the limits range. This technique is most commonly used in mechanical, electronics, and chemical industries where components or factors are assembled which increase the connection or involve structural subtraction. Also, this technique is used in computer simulation for determining optimal tolerances.

1.2.1.10.2   Advantages

Calculation of the total statistical tolerances is based on a single tolerance set of a total tolerance, which would be smaller than the overall estimate gained by arithmetic. Therefore, giving a general dimensional tolerance using wider tolerances is made possible with simpler and less costly production methods for single dimensions; this feature can be an important advantage in many cases.

1.2.1.10.3   Disadvantages

In general, the following prerequisites are needed for applying this technique to be feasible:

- Real single dimensions should be considered as non-correlated random variables
- The dimensional chain must be linear and have at least four members
- Single tolerances should be given in order of magnitude
- The distribution of single chain dimensions must be clear

If any of the above prerequisites are ignored, the analytical value of the application of this technique is lost. That is to say, the production of individual dimensions should be controlled and continuously monitored.

1.2.1.11   Time Series Analysis

The analysis of time series incorporates a set of methods for studying a batch of counting observations. This set includes:

- Punctuation of times series
- Finding delay patterns
- Finding periodic or seasonal patterns
- Forecasting future observations

1.2.1.11.1   Applications

Time series analysis is used to describe patterns of data in order to identify the set of points that should be reviewed. This technique is also used to create adjustments and to understand patterns of change besides specifying the points of change. The analysis is a technique used to predict future values (performance patterns over time) with the upper and lower limits defined as the prediction distance. As a result, the method can be applied to a vast range of time-consuming processes. For example, the issue of supplier assessment, the process of complaints by customers, the forecast of the procurement or repair of parts and related costs, the estimation of future energy consumption for production, and service collections are among the most important uses of the technique in question. This technique is used to set the process toward targets with the least variables. An example of the use of this method is in selecting and working with the desired quality suppliers (of an appropriate assessment score) so that the product's quality is maintained and/or a certain level of product supply is targeted to retain the market demand.

1.2.1.11.2   Advantages

Analyzing time series is most useful in the following cases: Planning, controlling engineering, identifying process changes, creating predictions, comparing planned performances of a process with certain criteria, and measuring the effect of some interventions or external operations. This technique provides an insight into causative patterns, and can be used further to distinguish systematic causes from specific causes. Another application of this method is to aid understanding of how a process would behave under certain conditions, and what settings would be needed for its effective implementation.

1.2.1.11.3   Disadvantages

Different techniques used to estimate time series, depending on the number of periods considered for the data, yield different responses. Add to this the nature of the data, the purpose, and the characteristics of the analysis, and the related costs should also be taken into account to achieve the desired result. Otherwise, the obtained results would be misleading.

1.2.1.11.4   A Summary of the Direct Application of the Statistical Techniques in Some Industries During the Last Decade

There are many examples of statistical techniques being used in industrial applications; they are described below.

In a comprehensive research study, design of experiments (DOE) was introduced as a powerful statistical technique for collecting QET and statistical tools. The authors of the study found that the desirable features in the integrity of processes are indicated for organizations that invoke ISO9001 and concentrate on the characteristics and trends of processes and products for attaining the measurement results. In other words, the innovation of this research lay in the application of DOE to measure the impacts of various factors on the process of a quality management system. In this study, the authors investigated the history of applications and the benefits of DOE, and the way the results are analyzed (Karbasian and Rostamkhani, 2017).

Innovative research revealed that the statistical process control (SPC) proves very effective in relation to the productivity calculations of construction companies (Espinosa-Garza et al., 2017).

In a valuable research project, the industrial production losses were fully assessed through DOE, SPC, and process capability indices (PCI) (Bounazef et al., 2014).

Research at the Malek Ashtar University of Technology in Iran, introduced statistical techniques as a tool for the mathematical branch of quality engineering. The statistical techniques were considered and applied by executive managers of quality management systems. Choosing these techniques and their application were entirely determined by the level of the organization's performance, where the requirements and recommendations of the quality management system are taken into account. Certainly, using a statistical techniques approach in quality engineering is a powerful, advantageous, and practical tool as it utilizes both the scientific aspect and the modern framework leading to organizational productivity. In the concluding part of their research, they explained the crucial and major benefits of statistical techniques as follows (Karimi Gavareshki et al., 2014):

- Finding the root cause of problems
- Quick solution to qualitative problems
- Augmenting customer satisfaction
- Attaining sustainability and capability in quality control processes
- A profitable tool for continuous improvement
- Creating awareness of qualitative situations, observation, and follow-up
- Creating data from quality management process
- Sustainable development of quality
- Supporting regular quality measurements and eliminating previous problems
- Developing existing products or processes
- Standardization and verification of all processes

In the aforementioned research, for the strategic issues related to ISO9001, at least ten key indices are defined, and for these ten indices, ten effective statistical techniques are considered as input variables. This forms the main basis in statistical analyses for both the working procedures of the Defense Industries Organization (DIO) and the performance of the Maham Group (MG) in Iran. The detailed functional model is presented as a final result of this study in Table 1.1.

In another research project, the generalized application of reliability concepts to outsourced supply chain networks was investigated. The authors introduced a new model in their research dealing with manufacturing lines with reworking and multiple parallel approaches, the results of which can be generalized to outsourced supply networks. Further, the results of this study, intending to ensure the optimized arrangement of outsourced supply chain networks, show the technique can be used to create a strong decision-making process for high-productivity manufacturing (Abbasi and Rostamkhani, 2014).

TABLE 1.1
Detailed Functional Model for the Effective Employment of Statistical Techniques in QMS[a]

Related statistical techniques	Related indices	Strategic issues
Simple bar charts Pareto charts Dispersion charts Statistical hypothesis tests Sampling and regression Design and analysis of experiments Reliability analysis Statistical process control charts Time series analysis	1. Getting customer feedback 2. Analyzing, investigating, and finding root causes of customer complaints (prioritizing corrective actions in early stages) 3. Analysis of customer satisfaction in the context of cause and effect relationship (to measure the effect of a particular cause on customer satisfaction). 4. Evaluation of the average customer satisfaction score Phase 1. Estimated sample number (customer) Phase 2. Investigating the organization's claims in the average customer satisfaction rating Phase 3. Measuring the effectiveness of corrective actions to increase customer satisfaction toward strengthening the above measures, and proceeding toward TQM[b]	Increasing customers' satisfaction
Process capability analysis Design and analysis of experiments Reliability analysis Statistical process control charts Time series analysis Sampling and regression Statistical tolerances	Conformity to product requirements in order of priority: 1. Observing customers' requirements, in an explicit or implicit manner 2. Compliance with the design requirements of the organization, including design standards 3. Compliance with health, safety, and environmental regulations (HSE)[c] 4. Compliance with state laws 5. Other agreement requirements	Product conformity
Statistical process control charts Process capability analysis Independent hypothesis test Design and analysis of experiments Histogram charts Time series analysis Sampling and regression	Phase 1. Assessing process trends in terms of being controlled or not. (confirming or not confirming the initial operation of the process) Phase 2. Checking the status of output processes under controlled conditions (to determine the desirability of the output of controlled processes) Phase 3. Measuring the nature of the process performance even under controlled conditions as well as desirability of their output Phase 4. Analysis of the relationship between different factors (such as periods, etc.) as sources of turbulence with process performance	Assessing and analyzing features of processes and products

(Continued)

TABLE 1.1 (CONTINUED)

Detailed Functional Model for the Effective Employment of Statistical Techniques in QMS[a]

Related statistical techniques	Related indices	Strategic issues
	Supplier status assessment includes:	
	Phase 1. Initial identification of suppliers	
Time series analysis	Phase 2. Choosing top suppliers	Reinforcement of suppliers
Histogram and trend charts	Phase 3. Periodic control of suppliers	
Reliability analysis	Phase 4. Identifying weaknesses and strengths of	
Sampling and regression	suppliers	
	Phase 5. Development and improvement of suppliers' capacity	

[a] Quality Management System
[b] Total Quality Management
[c] Health, Safety, and Environment

1.2.2 The History of Research in Non-Statistical Techniques

There are many non-statistical techniques that do not fall into a specific category. In the following section, however, the important and applicable non-statistical techniques are introduced as:

- Quality function deployment (QFD)
- Value engineering (VE)

QFD, a customer-focused approach for designing and improving product quality—the most common implementation tool in this technique—contains the following four matrices:

- Product planning matrix
- Product design matrix
- Process design matrix
- Process control matrix

Analysis of value function is the essence of VE aiming at identifying profitable areas for future studies. According to the definition provided in the standard EN12973:2000, function analysis reflects the impact of a product or implementation of a product. Also, this function analysis includes performance identification with expressing performances with logical tools. In this research, after a comprehensive overview of the history of important applications of QFD and VE techniques, non-statistical techniques were found to be the most significant techniques in quality engineering from 2008 to 2015. We have provided a table at the end of this review which makes the literature review in this regard a satisfactory one. Over recent years, we have witnessed individual applications from QFD and VE based on Lean and agile models in industrial companies. In this research, the combinatorial QFD and VE following a Lean approach as research methodology was explained. In other words, it is for the first time that the power of these non-statistical techniques is being illustrated in the form of integrated situations for control tests in product design (Karimi Gavareshki et al., 2017).

The value stream mapping (VSM) process allows one to create a detailed visualization of all steps in the work process. It is the representation of the flow of goods from a supplier to a customer through each organization. The primary purpose of creating a value stream map is to show the places where we can improve our process by visualizing both its value-adding and wasteful steps.

Workflow analysis (WFA) is a review of all sub-processes related to a specific operation. It can include plans to eliminate the inefficiencies and to optimize the efficiency of sub-processes.

Many industrial companies and firms have integrated strategic management tools with non-statistical techniques. The high flexibility of non-statistical techniques, for instance the fact that they can be developed, has convinced a majority of experts in various fields of management to apply them in their organizations (Rezazadeh et al., 2017).

The simplicity and accessibility of non-statistical techniques have encouraged their use by many experts in the economics and business sectors (Zhou, 2016). Moreover, the capability of non-statistical techniques to be used for interpreting such interdisciplinary issues as project management (PM) has created a strong background for analyzing different aspects of PM (Fisher, 2014).

1.3 THE HISTORY OF RESEARCH IN THE KEY CONCEPTS OF CONTINUOUS IMPROVEMENT

There are triple factors to achieving continuous improvement in industries that are presented in Figure 1.3.

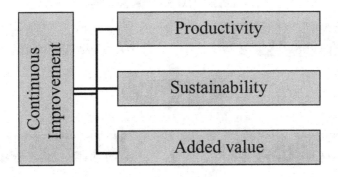

FIGURE 1.3 Triple factors for continuous improvement.

1.3.1 The History of Research in Productivity for Industrial Products Manufacturing

The best definition given for productivity divides the concept into two distinct phenomena, one is qualitative and the other one is quantitative in nature: i.e., availability and profitability. Of course, in a great number of industries or firms, the combination of efficiency and effectiveness has been offered as the definition of productivity (Plag, 2006). Customers nowadays judge organizations more and more on the basis of "how" these organizations have produced the products on offer. This "how" applies to a wide range of factors affecting the overall quality of the process including not only the quality of the goods/processes but also the quality in relation to the environment, workers, and ethical standards considered in terms of productivity. Some countries, like New Zealand, Australia, France, Holland, Denmark, Spain, and Great Britain have developed and started to draft national standards on Integrated Management Systems (IMS). However, there are some elements such as structure, size, and economic sector that may play a decisive role in influencing whether an organization decides to integrate systems and what are the breadth or depth of this integration. The comprehensive research in Italy has illustrated that the firms' desire to reduce the costs and time has compelled them to manage the triple systems of ISO9001, ISO14001, and OHSAS 18001. This means that the firms in the countries named above, especially in Italy have a strong desire toward productivity in manufacturing industrial products (Salomone, 2008). A research study has investigated the important role of some of the main domains of industrial engineering for productivity, for the first time. A schematic representation of the findings of this research project is given in Figure 1.4 (Karimi Gavareshki et al., 2018).

In this research project, the percentage of the impact of QET on productivity has been determined to be at 69.00%.

FIGURE 1.4   Five key factors in industrial engineering field that have impacts on productivity.

1.3.2 THE HISTORY OF RESEARCH IN SUSTAINABILITY FOR INDUSTRIAL PRODUCTS MANUFACTURING

The best definition of sustainability is the one concerned with confining human activities within the carrying capacity of the ecosystem (e.g., materials, energy, land, and water, etc.) that prevail in a locality placing emphasis on the quality of human life (air quality, human health). Moreover, economic sustainability considers the efficient use of resources to enhance operational profits and maximize market values. It also deals with substituting natural resources for manmade ones, reusing, and recycling. However, social sustainability focuses on the social well-being of the populace, balancing the need of an individual with the need for the group (equity), public awareness, and cohesion as well as participation and utilization of local labor and firms. It is acknowledged that the approach to sustainability differs depending on the field of application: Engineering, management, ecology, etc., to give some examples. Sustainability assessment is an appraisal method for evaluating the implementation level of sustainability measures. The sustainability assessment results are to be used for decision-making and policy formulation in real-world applications (Olawumi and Chan, 2018).

To deal with the increasing complexities of industries, a set of new methods have been developed to overcome the limitations of traditional methods. The statistical and stochastic approaches to production systems, the forecasting, heuristic and structural equations models, the black box, gray box and leveling methods, the fuzzy networks and maturity models are some of the new generation "tools" available to managers for retrieving information from what seemed to be a chaotic and impenetrable field of research. A brief revision of the literature shows the external applicability of maturity models. Specifically concerning organizational issues and quality management systems (QMS), it should be mentioned that is the one maturity approach adopted by the European Foundation for Quality Management Models. A potential maturity assessment guide on externalities is presented in a table where successful sustainability forms one of its most important ingredients. It can be concluded that establishing the maturity model has brought about the sustainability of industrial products manufacturing (Domingues et al., 2016).

The important role of some of the main domains of industrial engineering in sustainability has been studied for the first time in the research project presented in Figure 1.5 (Karimi Gavareshki et al., 2018).

The percentage of the impact of QET on sustainability in this research project has been determined to be at 69.44%.

1.3.3 THE HISTORY OF RESEARCH IN ADDED VALUES FOR INDUSTRIAL PRODUCTS MANUFACTURING

Over the recent decades, the issues concerning added values in business and economics sectors have been investigated. Added value analysis in business processes focuses on surveying the costs of the entire value chain, for example from receiving product orders to dispatching the output to the customer. The analysis strictly differentiates between the costs incurred for generating added values and costs of

FIGURE 1.5 Four key factors in industrial engineering field that have impacts on sustainability.

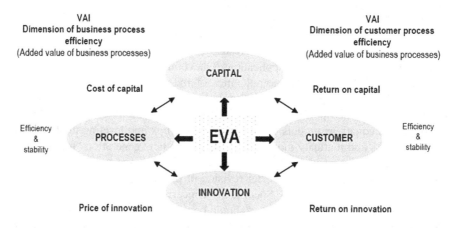

FIGURE 1.6 Relationship between EVA and BPVA.

non-productive procedure e.g., non-value-added activities and processes. In other words, economic value added (EVA) primarily concentrates on the costs of capital employed. While business processes value added (BPVA) focuses on the cost of complex value chain continuous time. The relationship between EVA and BPVA is presented in Figure 1.6 (Rajnoha et al., 2012).

1.4   A SUMMARY OF THE RESEARCH HISTORY

As can be seen in the literature research section of this book, there are many types of researches dealing with applying QET, both statistical and non-statistical. Moreover, the concepts of productivity or sustainability based on added values has been investigated seriously, especially in the domains of economic issues and business processes.

1.4.1   INNOVATION AND CREATIVITY IN THIS RESEARCH: A DESCRIPTION

The first and foremost innovation of this research is establishing a process-oriented model of QET in three types of processes in industrial products manufacturing. The appropriate statistical and non-statistical techniques are also observed in each of these processes. The second innovation is the definitions furnished for action plans for developing added values in relevant processes. The third innovation is establishing a verification path for this model in the determined scope. The fourth innovation is the calculations made of the total score of productivity and sustainability for the industry under study (determined scope) both before and after implementing the model.

EXERCISES

1.1. What are the main specifications of QET that should be considered by researchers? Can you form an assessment matrix between QET and productivity/sustainability?

1.2. What are the main domains of supply chain management (SCM) that might be generally considered? Can you form an assessment matrix between SCM and productivity/sustainability?

2 Applied Methodology

2.1 INTRODUCTION

The materials and methods used in the theoretical section of the research consist of data collection tools, reference books, and articles as well as written reports issued by the Iranian Centre for Defense Standards. The method for practical research that follows assessed the manufacturing industries belonging to some selected industries of the Defense Industries Organization (DIO). Moreover, questionnaires were used and administered among representative managers of the DIO. Exploratory interviews were also conducted.

2.2 RESEARCH LIMITATIONS

The protective measures enforced by defense organizations for security or other reasons are an indispensable part of their procedures. This feature has always been a limiting factor in the flow of information in the defense departments even among its own staff. That is, the classified nature of the information is strictly preserved. That is why it was not possible to study the audit reports of the Iranian Centre for Standardization outside the DIO or other defense agencies and subsidiaries of the Ministry of Defense and Armed Forces Logistics.

2.3 STATISTICAL POPULATION OF THE RESEARCH

The statistical population of this research is made up of some selected industries affiliated with the DIO. The quantitative characteristics of the statistical society are as follows:

- The sample size of the statistical population is 26 persons selected from management representatives of different industries
- Significant level = 0.05
- Test power = 80%
- The average effect volume = 50%

It should be noted that this productive and sustainable model of quality engineering techniques (QET) has been deployed in one industry, but, for assessing the model, our questionnaires were distributed among the 26 management representatives of the industries related to the DIO. In this study the following formula is used to determine the sample size required for the statistical assumptions on the mean value of the statistical population that follows (see Table 2.1).

$$Z_\beta = \frac{d(n-1)n^{1.2}}{(n-1)+1.21(Z_\alpha - 1.06)} - Z_\alpha \qquad (2.1)$$

The following figures show the personal characteristics of the statistical population (see Figures 2.1 to 2.6).

TABLE 2.1
Formula Description

Symbols	Description
n	n-sample size
d	The volume of the effect
z_a	Coefficient corresponding to a significant level of α in the distribution to the norm
z_β	Coefficient corresponding to the probability of the second type error β

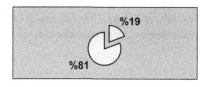

Master	19%
Bachelor	81%

FIGURE 2.1   Distribution diagram of a statistical sample in terms of education.

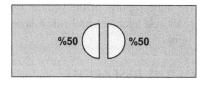

Manager	50%
Expert	50%

FIGURE 2.2   Distribution diagram of a statistical sample in terms of organizational positions.

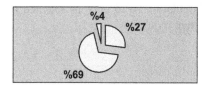

Less than 10 years	27%
Between 10 and 20 years	69%
More than 20 years	4%

FIGURE 2.3   Distribution diagram of a statistical sample in terms of work experience in quality fields.

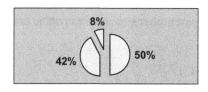

Auditor	50%
Lead auditor	42%
Trainee	8%

FIGURE 2.4   Distribution diagram of a statistical sample in terms of familiarity with quality systems.

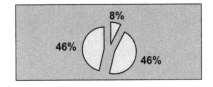

FIGURE 2.5   Distribution diagram of a statistical sample in terms of familiarity with QET.

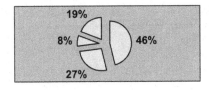

FIGURE 2.6   Distribution diagram of a statistical sample in terms of product volume.

2.4   DATA COLLECTION TOOLS AND METHODS

Questionnaires were administered among the 26 representative managers of the DIO. Exploratory interviews were also included.

2.5   DATA COLLECTION VALIDITY

The validity of this research was ascertained drawing upon expert opinion (including that of industrial advisers, lead auditors of the Iranian Center for Defense Standards, and university professors).

2.6   DATA COLLECTION RELIABILITY

To determine the reliability of this research, the Cronbach's alpha coefficient was used.

2.7   INFORMATION ANALYSIS METHOD

The data was analyzed through questionnaires and the reliability was measured using Cronbach's alpha coefficient and SPSS software. However, in order to prepare appropriate and applicable statistical tables, we have exploited Excel software and Minitab applications.

EXERCISE

2.1. What exactly is the difference between validity and reliability in any research?

3 Proposed Model in Triple Organizational Processes

3.1 INTRODUCTION

This chapter presents the findings of the research. We are concerned with traditional processes in the majority of organizations. The main proposed model of quality engineering techniques (QET) is introduced in the section related to all processes accompanied by instructive detailed examples of these applications.

The needed formulas related to all statistical calculations have been extracted from two brilliant references (main reference books) that have been used by many scholars (Montgomery, 1996a) (Montgomery, 1996b).

3.2 IDENTIFICATION AND DETERMINATION OF THREE TYPES OF PROCESSES IN INDUSTRIAL FACTORIES

Traditionally, there are triple processes in the industrial factories of Defense Industries Organization (DIO) in Iran. They are as follows:

- Main processes
- Leadership processes
- Support processes

Main processes in Quality Management Systems (QMS) based on ISO9001 play a vital role in meeting the basic requirements expressed in relevant standards.

Leadership processes in Quality Management Systems (QMS) based on ISO9001 play an important role in achieving organizational goals. These processes have the key responsibilities of managing and leading all defined processes.

Support processes in Quality Management Systems (QMS) based on ISO9001 have a supportive role in all processes belonging to the different levels of each organization.

These three types of processes are displayed in Figure 3.1 (based on ISO9001).

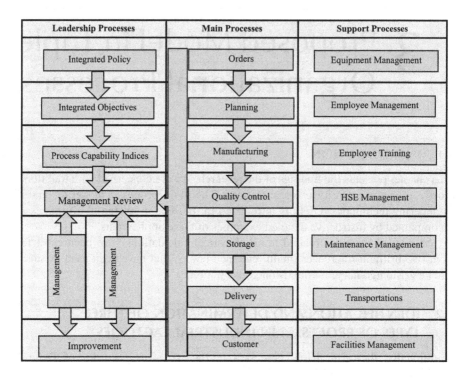

FIGURE 3.1   Three types of processes for industrial factories in DIO.

3.3   IDENTIFICATION AND DETERMINATION OF THE DESIRED QET IN MAIN PROCESSES

These processes divided into relevant sub-processes based on the suggested QET are exhibited in Figure 3.2.

3.3.1   A NUMERICAL APPLICATION OF QET IN THE DETERMINED SCOPE FOR MAIN PROCESSES

- Product design (*Quality Function Deployment + Value Engineering*)

Bullet-head drills with 27.8-inch HSD, powerful explosive devices in which immediate and concentrated release of energy are stored can penetrate oil and natural gas. Intense pressure from the explosion creates holes in the wall of the well from which oil and gas flow out. These types of bullets contain no caps and are fired by authorized participants. The main specifications that meet customers' requirements are presented in Table 3.1 (Karimi Gavareshki et al., 2017).

In this section, the proposed model for a bullet-head drill of 27.8-inch HSD is determined in triple matrices of quality function deployment (QFD).

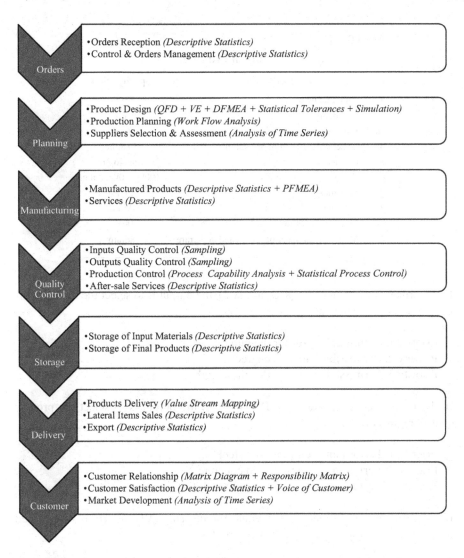

Orders
- Orders Reception *(Descriptive Statistics)*
- Control & Orders Management *(Descriptive Statistics)*

Planning
- Product Design *(QFD + VE + DFMEA + Statistical Tolerances + Simulation)*
- Production Planning *(Work Flow Analysis)*
- Suppliers Selection & Assessment *(Analysis of Time Series)*

Manufacturing
- Manufactured Products *(Descriptive Statistics + PFMEA)*
- Services *(Descriptive Statistics)*

Quality Control
- Inputs Quality Control *(Sampling)*
- Outputs Quality Control *(Sampling)*
- Production Control *(Process Capability Analysis + Statistical Process Control)*
- After-sale Services *(Descriptive Statistics)*

Storage
- Storage of Input Materials *(Descriptive Statistics)*
- Storage of Final Products *(Descriptive Statistics)*

Delivery
- Products Delivery *(Value Stream Mapping)*
- Lateral Items Sales *(Descriptive Statistics)*
- Export *(Descriptive Statistics)*

Customer
- Customer Relationship *(Matrix Diagram + Responsibility Matrix)*
- Customer Satisfaction *(Descriptive Statistics + Voice of Customer)*
- Market Development *(Analysis of Time Series)*

FIGURE 3.2 Desired QET in main processes.

In the product design matrix (QFD matrix 1), the output is to select the product quantitative specifications:

Priority 1: Minimum diameter of arrival and departure
Priority 2: Minimum pressure
Priority 3: Maximum safe stretch
Priority 4: Gun degree
Priority 5: Gun diameter

TABLE 3.1
Specifications of Technical Product

N	Description	Specification
1	Gun thickness	73mm
2	Gun degree	60
3	Bullet numbers per unit	6spf
4	Bullet distances from one another	50.8mm
5	Gun length	1.5m–6.3m
6	Permissible gun weight	100kg, 50kg, 25kg
7	Maximum pressure	25kpsi
8	Maximum safe stretch	190kIb
9	Minimum pressure	0bars
10	Minimum diameter of arrival and departure	86mm

In the process design matrix (QFD matrix 2), the output is to select the manufacturing process requirements:

Priority 1: Permissible limit for the gun diameter
Priority 2: Permissible limit for degree
Priority 3: Permissible limit for the gun length

In the process control matrix (QFD matrix 3), the output is to select the process control requirements:

Priority 1: Performance on proposed steel
Priority 2: Performance on proposed concrete
Priority 3: Performance at a temperature of 190 °C

Finally, the proposed model in the Value Engineering (VE) matrix is determined together with prioritizing the product control tests:

Priority 1: Performance test at a temperature of 190 °C
Priority 2: Performance test on proposed steel
Priority 3: Performance test on proposed concrete

It should be noted that in the proposed model, we have used three matrices with four possible matrices in QFD.

Also, our focus is on the row requirements over the column requirements at each matrix in QFD. This means that the points in column (E) in each matrix are multiplied by the points in the same column. In fact, the reason why these points in the same column are selected is that the column requirements are satisfied such that the numbers between 1, 3, and 9 are chosen. The results of the proposed model can be

QFD Matrix 1	Product's quantitative specifications										QFD parameters						
Bullet-head drill, 27.8inch HSD	Gun diameter=73mm	Gun degree=60	Bullet numbers per unit=6spf	Bullet distances from one another=50.8mm	Gun length=1.5m-6.3m	Permissible gun weight=100kg-50kg-25kg	Maximum pressure=25kpsi	Maximum safe stretch=190klb	Minimum pressure=0bar	Minimum diameter of arrival and departure=86mm	A (Importance degree)	N (Organization assessment)	P (Organization plan)	B (Improving ratio)	C (Correction factor)	D (Absolute weight)	E (Relative weight)
Suitable performance of product	9	9	9	3	3	1	1	3	3	9	5	5	5	1	1.5	7.5	24
Good appearance of product	1	1	1	9	9	9	9	1	1	1	3	2	4	2	1.2	7.2	23
Product reliability	3	3	3	1	1	1	1	9	9	9	4	3	5	1.7	1.5	10.2	33
After sale services	3	3	1	1	1	1	1	1	1	1	4	3	4	1.3	1.2	6.2	20
Absolute weight	398	398	358	332	332	284	284	412	412	556	3766						
Relative weight	10.6^5	10.6^4	9.5	8.8	8.8	7.5	7.5	10.9^3	10.9^2	14.8^1	100			Total		31.1	100

(Customer qualitative requirements)

FIGURE 3.3　Product design matrix.

seen in Figures 3.3 to 3.6 (Karimi Gavareshki et al., 2017). The relevant formulas are as follows (3.1):

$$B = \frac{P}{N}$$

B: Improving ratio
P: Organization plan
N: Organization assessment

$$D = A \times B \times C$$

QFD Matrix 2	Manufacturing process requirements						QFD parameters						
Bullet-head drill, 27.8inch HSD	Permissible limit for gun diameter =±0.5mm	Permissible limit for degree =±0.5	Permissible limit for bullet distances =±0.5mm	Permissible limit for gun length =±0.1m	Permissible limit for gun weights =±0.5kg	Permissible limit for Pressure =±0.5kpsi	A (Importance degree)	N (Organization assessment)	P (Organization plan)	B (Improving ratio)	C (Correction factor)	D (Absolute weight)	E (Relative weight)
Product's quantitative specifications													
Minimum diameter of arrival and departure = 86mm	1	1	1	1	1	1	5	5	5	1	1.5	7.5	20
Minimum pressure = 0bar	1	1	1	1	1	1	3	2	4	2	1.2	7.2	19
Maximum safe stretch = 190kIb	1	1	1	3	1	1	4	3	4	1.3	1.5	7.8	21
Gun degree = 60	3	9	1	3	3	1	5	5	5	1	1.5	7.5	20
Gun diameter = 73mm	9	3	1	3	3	1	5	5	5	1	1.5	7.5	20
Absolute weight	300	300	100	222	180	100	1202	Total				37.5	100
Relative weight	25[1]	25[2]	8.3	18.5[3]	15	8.3	100						100

FIGURE 3.4   Process design matrix.

D: Absolute weight
A: Importance degree
B: Improving ratio
C: Correction factor

$$E = \frac{D}{T}$$

E: Relative weight
D: Absolute weight

$$T = \sum_{i=1}^{n} D_i$$

QFD Matrix 3	Process control requirements					QFD parameters						
Bullet-head drill, 27.8inch HSD	Bullet inflicting no damage, no injury, no scratch	Meeting packaging requirements	Performance on object steel according to instructions No.1	Performance on object concrete according to instructions No.2	Performance at a temperature of 190 °C according to instructions No.3	A (Importance degree)	N (Organization assessment)	P (Organization plan)	B (Improving ratio)	C (Correction factor)	D (Absolute weight)	E (Relative weight)
Permissible limit for gun diameter = ±0.5mm	1	1	9	9	3	5	5	5	1	1.5	7.5	35
Permissible limit for degree = ±0.5	1	1	9	9	3	5	5	5	1	1.5	7.5	35
Permissible limit for gun length = ±0.1m	1	1	9	9	9	4	3	4	1.3	1.2	6.2	30
Absolute weight	100	100	900	900	480	2480						
Relative weight	4	4	36.3^1	36.3^2	19.3^3	100			Total		21.2	100

Manufacturing process requirements

FIGURE 3.5 Process control matrix.

- Suppliers Selection and Assessment (*Time Series Analysis*)

In a selected industry affiliated with DIO, a supplier assessment was carried out by a supervisor on site using a checklist prepared and arranged on a scale of 1000 points for a period of 9 years. The results are shown in Tables 3.2 to 3.4.

$$X = x - 82 \rightarrow \text{Changing of the variable}$$

$$b = \frac{\sum_{i=1}^{n} X_i Y_i}{\sum_{i=1}^{n} X_i^2} \tag{3.2}$$

| Value engineering in product control tests based on standard EN12973:2000 | | | | | | | | | | |
| Main specifications in product control tests | | | | | | | | | | |
Bullet-head drill, 27.8inch HSD		Ease of training and testing	Impact of testing on main function of product	Correlation with other control tests	Safety testing for person and environment	Minimum environmental considerations	Total columns need	Importance of testing (I=N×A)	Test cost per unit (Rials of I.R.I)	Test cost C (normalized)	Test Value (Value=I/C)
	Need	0.15	0.50	0.20	0.10	0.05	1				
product control tests headings	Performance test on object steel	4	5	5	3	2		4.5	750000	0.42	10.71^2
	Performance test on object concrete	4	4	4	3	1		3.75	650000	0.36	10.42^3
	Performance test at temperature of 190 °C	4	3	3	4	4		3.3	400000	0.22	15^1
	Total								1800000	1	36.13

FIGURE 3.6 Value engineering matrix.

In this formula, b is the angle coefficient of the line equation. So, we have:

$$b = 16.67$$

$$a = \bar{Y} = \frac{\sum_{i=1}^{n} Y_i}{n} \tag{3.3}$$

In this formula, (a) is the distance from (0, 0) on the Y-axis and (n) is the number of data or samples. Therefore, we have:

$$a = 740$$

$$\bar{Y} = a + b\bar{X} \tag{3.4}$$

TABLE 3.2
Supplier Data of Organization

N	X based on year	Y based on score (out of 1000)
1	2007	650
2	2008	700
3	2009	720
4	2010	750
5	2011	730
6	2012	710
7	2013	800
8	2014	820
9	2015	780

TABLE 3.3
Calculated Supplier Data (Three Years Moving Average)

N	X based on year	Y based on score (out of 1000)	Sum of 3 years	Average of 3 years
1	2007	650		
2	2008	700	2070	690
3	2009	720	2170	723
4	2010	750	2200	733
5	2011	730	2190	730
6	2012	710	2240	747
7	2013	800	2330	777
8	2014	820	2400	800
9	2015	780		

TABLE 3.4
Calculated Supplier data of Organization (Squares Minimum Method)

N	X based on year	Y based on score (out of 1000)	X_i	X_i^2	X_iY_i
1	2007	650	−4	16	−2600
2	2008	700	−3	9	−2100
3	2009	720	−2	4	−1440
4	2010	750	−1	1	−750
5	2011	730	0	0	0
6	2012	710	1	1	710
7	2013	800	2	4	1600
8	2014	820	3	9	2460
9	2015	780	4	16	3120
S		6660	0	0	1000

As a result, we can write the equation related to the above line as follows:

$$\overline{Y} = 740 + 16.67\overline{X}$$

To show the general equation related to the suppliers, we give for the variable X, two values such as ± 2

$X = \pm 2$ and with calculating the corresponding values for Y:

$$X = 2 \rightarrow Y = 740 + 16.67 \times 2 = 773.34 \approx 773$$
$$X = -2 \rightarrow Y = 740 - 16.67 \times 2 = 706.66 \sim 707$$

To predict the supplier's score in the year 2016, in exchange for variable X, we place 5. So, we have:

$$X = 5 \rightarrow Y = 740 - 16.67 \times 5 = 823.35$$

Figure 3.7 shows the supplier's score chart obtained by the squares minimum method.

- Production Control (*Process Capability Analysis*)

Appropriate specifications for a piece are set at 2.05 ± 0.02. If the size of the piece is less than the bottom limit of the specification, it is rejected; and if it exceeds the upper limit of the specification, it is corrected. Process Capability Analysis has been

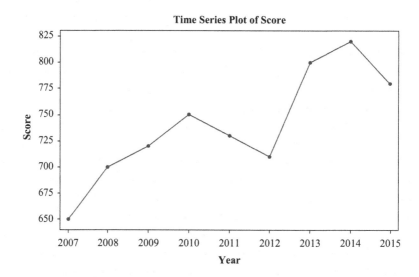

FIGURE 3.7 Supplier's score chart by squares minimum method.

used to control the production of this piece. Assuming that the distribution of the production process is normal and under statistical control, the following results are obtained:

$$\sum_{i=1}^{k} \overline{X_i} = 41.25, \sum_{i=1}^{k} R_i = 0.32, k = 20, n = 4$$

In this equation statement, the mean $\overline{X_i}$ is in a subgroup, R_i is the subgroup range, k is the number of subgroups, and n is the number of subgroup members.

1. Determine the process standard deviation.

$$\overline{R} = \frac{\sum_{i=1}^{k} R_i}{k} \tag{3.5}$$

So, we have: $\overline{R} = 0.016$
Moreover, we have:

$$\hat{\sigma} = \frac{\overline{R}}{d_2} \tag{3.6}$$

$\hat{\sigma}$: Estimated standard deviation
d_2: Fixed coefficient related to the number of subgroup members
In this example, for the four members in the subgroup, we will have:

$$d_2 = 2.059 \rightarrow \hat{\sigma} = 0.0078$$

2. Determine the process capability index (potential) CP and the process capability index (actual) CPK.

$$\overline{\overline{X}} = \frac{\sum_{i=1}^{k} \overline{X_i}}{k}, \quad \overline{\overline{X}} \rightarrow \text{Total Mean Value of all subgroups} \tag{3.7}$$

So, we have:

$$\overline{\overline{X}} = 2.0625$$

Moreover, we have:

$$C_P = \frac{\text{USL} - \text{LSL}}{6\sigma} \tag{3.8}$$

USL is the upper limit and LSL is the lower limit of the piece tolerance. Therefore, we have:

$$C_P = \frac{2.07 - 2.03}{6 \times 0.0078} = 0.8547$$

Moreover, we have:

$$C_{PK} = \text{Min} \left\{ \frac{\text{USL} - \overline{\overline{X}}}{3\sigma}, \frac{\overline{\overline{X}} - \text{LSL}}{3\sigma} \right\} \qquad (3.9)$$

So, we have:

$$C_{PK} = \text{Min} \left\{ \frac{2.07 - 2.0625}{3 \times 0.0078}, \frac{2.0625 - 2.03}{3 \times 0.0078} \right\} = \text{Min} \{0.32, 1.39\} = 0.32$$

3. Calculate the standard deviation from the manufacturing process.

$$\delta = \frac{|m - \overline{x}|}{\dfrac{\text{USL} - \text{LSL}}{2}}, m = \frac{\text{USL} + \text{LSL}}{2} \qquad (3.10)$$

USL is the upper limit and LSL is the lower limit of the piece tolerance. Thus, we have:

$$m = 2.05, \delta = \frac{|2.05 - 2.0625|}{\dfrac{2.07 - 2.03}{2}} \rightarrow \delta = 0.6252$$

Also, with the formula in the form of

$$\left[C_{PK} = C_P (1 - \delta) \right] \qquad (3.11)$$

the same result is obtained.

4. What percentage of these pieces is rejected and what percentage needs correction?

$$P(X > 2.07) = P(Z > \frac{2.07 - 2.0625}{0.0078}) = P(Z > 0.96) \qquad \text{Need Correction}$$

$$= 1 - P(Z < 0.96) = 1 - 0.8315 = 0.1685 \rightarrow 16.85\%$$

$$P(X < 2.03) = P\left(Z < \frac{2.03 - 2.0625}{0.0078} \right) = P(Z < -4.17) \approx 0 \rightarrow 0\% \text{ Rejected}$$

- Production Control (*Statistical Process Control for variable data*)

Data on the internal diameter measurements of a sample series related to special project pieces are given in Table 3.5 (nominal diameter of the piece is 1.51 ± 0.33).

1. What method of process analysis do you use?
2. Define the control limits. (The upper and lower limits).
3. Is the manufacturing process under control?

1. $\overline{X}$ & R
2.1. The following formulas are used to determine the control limits of the mean value chart:

$$\overline{X} = \frac{\sum_{i=1}^{k} \overline{X_i}}{k} \tag{3.12}$$

TABLE 3.5
Internal Diameter Data of a Piece under Four Groups

k	X_1	X_2	X_3	X_4	$\overline{X_i}$	R_i
1	1.50	1.51	1.50	1.51	1.505	0.01
2	1.51	1.52	1.50	1.51	1.510	0.02
3	1.50	1.51	1.51	1.51	1.507	0.01
4	1.51	1.51	1.50	1.51	1.507	0.01
5	1.50	1.50	1.51	1.51	1.505	0.01
6	1.49	1.50	1.50	1.50	1.497	0.01
7	1.50	1.50	1.51	1.50	1.502	0.01
8	1.49	1.51	1.50	1.50	1.500	0.02
9	1.50	1.50	1.50	1.49	1.497	0.01
10	1.50	1.49	1.50	1.51	1.500	0.02
11	1.50	1.50	1.50	1.51	1.502	0.01
12	1.50	1.49	1.49	1.50	1.495	0.01
13	1.50	1.49	1.49	1.49	1.492	0.01
14	1.50	1.48	1.49	1.49	1.490	0.02
15	1.49	1.49	1.50	1.49	1.492	0.01
16	1.50	1.49	1.49	1.49	1.492	0.01
17	1.49	1.48	1.49	1.49	1.487	0.01
18	1.48	1.49	1.48	1.49	1.485	0.01
19	1.48	1.49	1.49	1.49	1.487	0.01
20	1.49	1.50	1.49	1.49	1.492	0.01
21	1.49	1.49	1.48	1.49	1.487	0.01
22	1.48	1.47	1.48	1.49	1.480	0.02
23	1.47	1.48	1.49	1.48	1.480	0.02
24	1.47	1.48	1.50	1.49	1.485	0.03
					35.876	0.32

In this formula, $\overline{\overline{X}}$ is the total mean value in all subgroups and k is the number of subgroups. So, we have:

$$\overline{\overline{X}} = 1.495$$

$$UCL_{\overline{X}} = \overline{\overline{X}} + A_2\overline{R} \qquad (3.13)$$

In this formula, $UCL_{\overline{X}}$ is the upper limit of the mean value control chart and A_2 is the fixed coefficient and $\overline{R}$ is the mean value of ranges. So, we have:

$$UCL_{\overline{X}} = 1.495 + 0729 \times 0.013 = 1.504 \rightarrow \text{Upper limit}$$

$$LCL_{\overline{X}} = \overline{\overline{X}} - A_2\overline{R} \qquad (3.14)$$

In this formula, $LCL_{\overline{X}}$ is the lower limit of the mean value control chart and A_2 is the fixed coefficient and $\overline{R}$ is the mean value of ranges. So, we have:

$$LCL_{\overline{X}} = 1.495 - 0729 \times 0.013 = 1.485 \rightarrow \text{Lower limit}$$

The value for A_2 can be obtained from the table of coefficients and the value of $\overline{R}$ from the table given in the next section.

2.2. The following formulas are used to determine the control limits related to the range chart:

$$\overline{R} = \frac{\sum_{i=1}^{k} R_i}{k} \qquad (3.15)$$

In this formula, $\overline{R}$ is the average of ranges in all subgroups and k is the number of subgroups. So, we have:

$$\overline{R} = 0.013$$

$$UCL_R = D_4\overline{R} \qquad (3.16)$$

In this formula, UCL_R is the upper limit of the range control chart and D_4 is the fixed coefficient and $\overline{R}$ is the average of ranges. So, we have:

$$UCL_R = 2.282 \times 0.013 = 0.0297$$

$$LCL_R = D_3\overline{R} \qquad (3.17)$$

In this formula, LCL_R is the lower limit of the range control chart and D_3 is the fixed coefficient and $\overline{R}$ is the average of ranges. So, we have:

$$LCL_R = 0 \times 0.13 = 0$$

Therefore:

The control limits of the mean value control chart are calculated as:

$$\begin{cases} UCL_{\bar{x}} = 1.504 \\ LCL_{\bar{x}} = 1.485 \end{cases}$$

The control limits of the range control chart are calculated as:

$$\begin{cases} UCL_R = 0.0297 \\ LCL_R = 0 \end{cases}$$

3. Analysis
   - Considering the data in Table 3.5, and comparing the data in columns of X_1 to X_4 with those of the control limits of the mean value control chart, note that our data is not under control. (As there are some points that are outside of these control limits.).
   - Considering the data in Table 3.5, and comparing the data in column of R_i with those of the control limits of the range control chart, note that our data is under control. (As there are no points that are outside of these control limits.)

So, in general, the manufacturing process related to these pieces is not under control.

The control charts on the mean value and range of the latter variable data are given in Figure 3.8.

As is shown, in the range chart, the data are under control, however, the mean value chart indicates that there are at least two types of out-of-control patterns.

- Production Control (*Statistical Process Control for descriptive data + Descriptive Statistics—Pareto Chart*)

Samples of four defects are detected in the inspection processes: Inhomogeneity, fragmentation, friction, and cracks. These defects are presented in Figure 3.9.

The data related to these four defects are provided in Table 3.6.

- What process analysis method do you use?
- Define control limits.
- Is the manufacturing process under control?
- Is the process capability index (actual) or C_{PK} calculable?
- Can you prioritize the corrective actions with Pareto's analysis?
  1. C technique
  2. The following formulas are used to determine the control limits of the C technique:

$$\bar{C} = \frac{\sum_{i=1}^{n} C_i}{n} \tag{3.18}$$

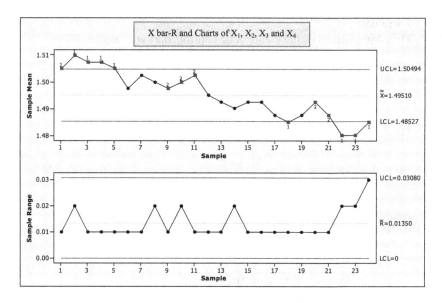

FIGURE 3.8 Control charts for mean value and range of variable data.

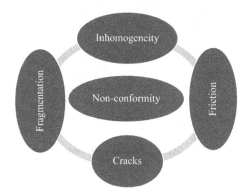

FIGURE 3.9 Non-conformity of four defects.

In this formula, $\overline{C}$ is the total average of non-conforming cases, and n is the total number of pieces. So, we have:

$$\overline{C} = 2.667$$

$$\mathrm{UCL}_C = \overline{C} + 3\sqrt{\overline{C}} \qquad (3.19)$$

In this formula, UCL_C is the upper limit of C chart, and $\overline{C}$ is the total average of non-conforming cases. So, we have:

$$\mathrm{UCL}_C = 2.667 + 3\sqrt{2.667} = 7.566$$

$$\mathrm{LCL}_C = \overline{C} - 3\sqrt{\overline{C}} \qquad (3.20)$$

TABLE 3.6

Number of the Non-Conformities in Project Pieces

N	Inhomogeneity	Fragmentation	Friction	Crack	C_i
1	0	1	0	1	2
2	0	1	2	0	3
3	0	2	0	0	2
4	0	1	0	0	1
5	0	1	0	1	2
6	1	2	0	0	3
7	1	2	0	0	3
8	0	1	1	1	3
9	0	1	2	0	3
10	1	2	0	0	3
11	0	1	1	0	3
12	1	0	1	0	2
13	0	1	2	0	3
14	1	1	1	0	3
15	1	2	1	0	4
16	1	1	0	1	3
17	1	2	1	0	4
18	1	2	1	0	4
19	0	1	1	0	2
20	0	2	0	0	2
21	1	0	1	0	2
22	0	2	0	1	3
23	0	1	2	0	3
24	0	0	1	1	2
S	10	30	18	6	64

In this formula, LCL_C is the lower limit of C chart, and $\bar{C}$ is the total average of non-conforming cases. So, we have:

$$LCL_C = 2.667 - 3\sqrt{2.667} = -2.232 \rightarrow 0$$

Therefore:

The control limits of C chart are calculated as:

$$\begin{cases} UCL_C = 7.566 \\ LCL_C = 0 \end{cases}$$

3. Analysis

Considering the data in Table 3.6 and comparing the data in the column of C_i with those of the control limits of the C chart, note that our data is under control (As there are no points that are outside of these control limits). The C chart is visible in Figure 3.10. As can be seen, the relevant chart shows the data that are under control.

4. The process being under control, we will have:

$$C_{PK} = \overline{C} = 2.667$$

5. Considering the four types of non-conformities, we have listed them in Table 3.7.

Figure 3.11 shows the Pareto chart of prioritizing corrective actions.

• Customer Satisfaction (*Descriptive Statistics—Pareto Chart*)

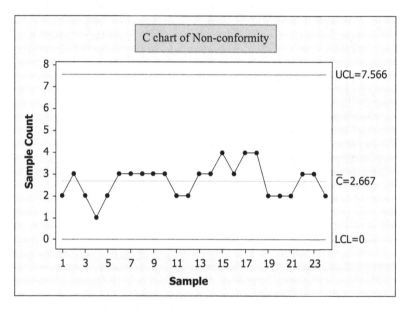

FIGURE 3.10   C control chart of non-conforming cases in project pieces.

TABLE 3.7

Registered Non-Conformities in the Project Pieces

N	Non-Conformity (defects)	Frequency
1	Inhomogeneity	10
2	Fragmentation	30
3	Friction	18
4	Crack	6
S		64

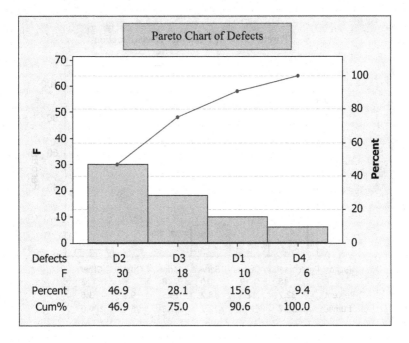

FIGURE 3.11   Pareto chart of prioritizing corrective actions.

In order to analyze and examine the causes of customer complaints in the selected scope in one industry in DIO during the one year, the information relevant to these complaints are extracted and presented in Table 3.8.

The Pareto chart in Figure 3.12 demonstrates the information relevant to Table 3.8 complete with the cumulative line on the chart.

The Pareto chart serves as a useful tool for prioritizing corrective actions to address customer dissatisfaction. Even so, it should be noted that in some cases the cause(s) of a problem in the organization might be interrelated and that it is not always easy to relegate a problem to a specific unit or department. Another point to remember is that, if Pareto charts are used to indicate the arrangement of the data

TABLE 3.8

Causes of Customer Dissatisfaction

N	Causes of dissatisfaction	Frequency
1	Dissatisfaction with product quality	5
2	Dissatisfaction with delivery time	12
3	Dissatisfaction with product quality (packaging)	18
4	Dissatisfaction with after-sales services	10
5	Dissatisfaction with price set for product	8
6	Dissatisfaction with employees' performance in organization	2
Total		55

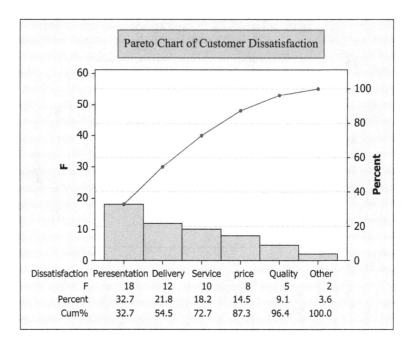

FIGURE 3.12   Pareto chart of customer dissatisfaction.

from the highest frequency to the lowest one for the determination of the causes related to the costs, most certainly this chart satisfies our needs to determine the highest cost items. However, there is a possibility that the second chart does not conform to the first one. For the causes related to the highest dissatisfactions do not always match causes related to the highest costs. The cumulative line shows the process slope related to causes.

- After-Sales Services and Customer Satisfaction (*Descriptive Statistics— Dispersion Chart*)

In the selected and determined scope in one industry in DIO, data on the relationship between the duration of after-sales services related to different key products and customer satisfaction (14 customers) are extracted and presented in Table 3.9.
    The relevant data is presented in Table 3.10.

$$b = \frac{n \sum_{i=1}^{n} X_i Y_i - \sum_{i=1}^{n} X_i \sum_{i=1}^{n} Y_i}{n \sum_{i=1}^{n} X_i^2 - \left(\sum_{i=1}^{n} X_i\right)^2} \qquad (3.21)$$

In this formula, b is the line angle coefficient and n is the number of data

So, we have: $b = \dfrac{(14 \times 190) - (61 \times 34)}{14 \times 351 - (61)^2} = 0.49$

TABLE 3.9

Score Model for Customer Satisfaction

Duration of after-sales services (in terms of years)	Satisfaction level (quantitative)	Satisfaction level (qualitative)
1	1	Low
3	2	Moderate
5	3	Good
7	4	Very good
9	5	Excellent

TABLE 3.10

Customer Satisfaction Data (after-sales services)

N	Duration of after-sales services (X in term of year)	Satisfaction level (quantitative) (Y in term of 1 to 5)	X^2	Y^2	XY
1	1	1	1	1	1
2	2	1	4	1	2
3	3	2	9	4	6
4	4	2	16	4	8
5	7	4	49	16	28
6	5	3	25	9	15
7	6	3	36	9	18
8	2	1	4	1	2
9	3	2	9	4	6
10	8	4	64	16	32
11	9	5	81	25	45
12	4	2	16	4	8
13	1	1	1	1	1
14	6	3	36	9	19
S	61	34	351	104	190

$$\overline{Y} = a + b\overline{X} \qquad (3.22)$$

Where a and b are fixed coefficients of the line equation.

So, we have: $a = \overline{Y} - b\overline{X} \rightarrow a = (34 \div 14) - 0.49 \times (61 \div 14) = 0.293$

As a result, the line equation is as follows:

$$\overline{Y} = 0.293 + 0.49\overline{X}$$

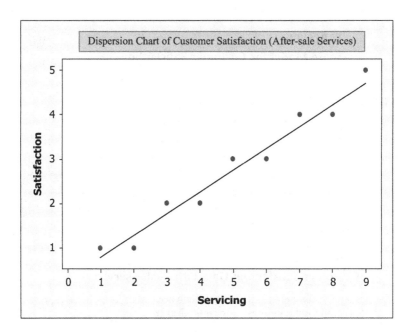

FIGURE 3.13 Dispersion chart of customer satisfaction (after-sales services).

Moreover, we have:

$$r = \frac{n\sum_{i=1}^{n} X_i Y_i - \sum_{i=1}^{n} X_i \sum_{i=1}^{n} Y_i}{\sqrt{\left[\left(n\sum_{i=1}^{n} X_i^2 - \left(\sum_{i=1}^{n} X_i\right)^2\right)\left(n\sum_{i=1}^{n} Y_i^2 - \left(\sum_{i=1}^{n} Y_i\right)^2\right)\right]}} \qquad (3.23)$$

In this formula, r is the correlation coefficient and n is the number of data

So, we have: $r = \dfrac{(14 \times 190) - (61 \times 34)}{\sqrt{[(14 \times 351 - (61)^2)(14 \times 104 - (34)^2)]}} \rightarrow R^2 = 0.9210$

The overall result of this calculation indicates that roughly 92% of the share related to customer satisfaction is derived from the index of after-sales services time. This shows the importance of the latter index in obtaining customer satisfaction. The dispersion diagram of the information given above, along with the line calculated is displayed in Figure 3.13.

The dispersion diagram in this numerical application has a linear and positive correlation pattern.

3.4 IDENTIFICATION AND DETERMINATION OF THE DESIRED QET IN LEADERSHIP PROCESSES

These processes divided into relevant sub-processes based on the suggested QET are presented in Figure 3.14.

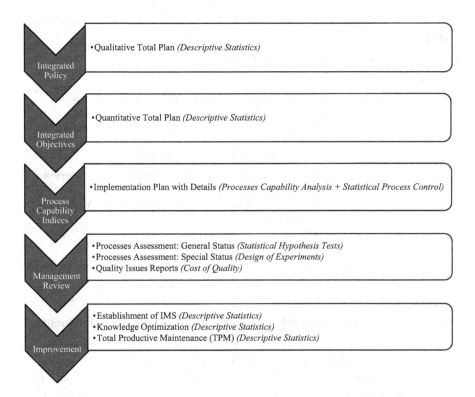

FIGURE 3.14 Desired QET in leadership processes.

Note that in our total process map, the productivity and sustainability processes are not independent entities. This interdependence is presented in Table 3.11.

3.4.1 NUMERICAL APPLICATION OF QET IN THE DETERMINED SCOPE FOR LEADERSHIP PROCESSES

- Processes Assessment: General Status (*Statistical Hypothesis Tests*)

The related indices of the selected industry in DIO are divided into three categories: the support, the main, and the leadership indices. To ascertain a confidence level of 95%, the performances of these processes indices are assessed as being independent or not. To that end, first, the performance table for the related industry with practical frequencies in different departments is set according to Table 3.12. Then, the performance table for the same industry with theoretical frequencies in the same departments is set according to Table 3.13.

The distribution function of χ^2 (K-Square) is obtained using the following formula:

$$\chi^2_{\alpha,df} = \sum_{i=1}^{m} \sum_{j=1}^{n} \frac{(Fe_{ij} - Fo_{ij})^2}{Fe_{ij}} \tag{3.24}$$

TABLE 3.11
Productivity and Sustainability in Processes

Processes	Sub-processes
Productivity management	Suppliers' assessment. Production productivity. Production control. Establishment of Information Management System (IMS).
	Industrial psychology. Work health. Work hygiene. Work safety. General trainings.
	Skills optimization. Knowledge optimization.
Sustainability management	Production sustainability. Product designing. Environmental hygiene (water + electricity + fuel) &
	(noise + air + pollutants). Mean time to failure (MTTF). Mean time to repair (MTTR).
	Availability (AV). Failure rate [R(t)]. Total productive maintenance (TPM).

TABLE 3.12
Performance of Indices in Selected Industry in DIO (Practical Frequency)

N	Department name	0 to 25%	25 to 50%	50 to 75%	75 to 100%	N_i
1	Top management	0	0	0	1	1
2	Quality assurance	1	2	5	10	17
3	Material and product planning	1	1	4	22	28
4	Internal and external business	1	2	4	16	23
5	Research and development	0	1	2	5	8
6	Financial and economics	0	0	2	4	6
7	Human resources	1	0	4	13	18
8	warehouses management	0	1	4	0	5
9	Safety and health	1	3	5	15	24
10	Manufacturing groups	2	2	8	10	22
	N_j	6	12	38	96	152

In this formula, Fe_{ij} is the theoretical frequency in the (i) row and (j) column and Fo_{ij} is the actual frequency in the (i) row and (j) column. It should be noted that df is the degree of freedom; and the related formula $df = (m - 1)(n - 1)$ is obtained when m is the number of rows in the table and n is the number of columns in the table. However, α is also the error or uncertainty. Moreover, we have:

$$Fe_{ij} = \frac{N_i \times N_j}{N} \tag{3.25}$$

In this formula, Fe_{ij} is the theoretical frequency in the (i) row and (j) column. N_i is the frequency in the (i) row and N_j is the frequency in the (j) column. N is the total number of frequencies.

TABLE 3.13

Performance of Indices in Selected Industry in DIO (Theoretical Frequency)

N	Department name	0 to 25%	25 to 50%	50 to 75%	75 to 100%	N_i
1	Top management	0	0	0	1	1
2	Quality assurance	1	1	4	11	17
3	Material and product planning	1	2	7	18	28
4	Internal and external business	1	2	6	14	23
5	Research and development	0	1	2	5	8
6	Financial and economics	0	0	2	4	6
7	Human resources	1	1	5	11	18
8	Warehouses management	0	0	1	3	4
9	Health and safety	1	2	6	15	24
10	Manufacturing groups	1	2	5	14	22
	N_j	6	11	40	96	

The test of the statistical independence χ^2 is as follows:

$\begin{cases} H.: \chi^2 \leq \chi^2_{df} & \text{In hypothesis } H., \text{ the performance of the process index is independent of the unit.} \\ H_1: \chi^2 > \chi^2_{df} & \text{In hypothesis } H_1, \text{ the performance of the process index is related to the unit.} \end{cases}$

(3.26)

So, we have:

$$\chi^2 = 1+1+\frac{1}{4}+\frac{1}{11}+\frac{1}{2}+\frac{9}{7}+\frac{4}{18}+\frac{4}{6}+\frac{4}{14}+1+\frac{1}{5}$$

$$+\frac{4}{11}+9+3+\frac{1}{2}+\frac{1}{5}+0+1+0+\frac{9}{5}+\frac{16}{14}$$

$$= 23.51$$

The calculations of the hypothesis test related to the χ^2 distribution function are as follows:

$$df = (m-1)(n-1) = 9 \times 3 = 27$$

$$\chi^2 > \chi^2_{0.05,27} \rightarrow \chi^2 > 4.01 \text{(critical area)} \rightarrow H. \text{ is rejected}$$

↓

The performances of the processes index are related to the units.

- Processes Assessment: Special Status (*Design of Experiments*)

The related indices of the selected industry in DIO are assessed in a classification, based on six-month periods as treatment data and subsidiary units as sources of disturbance data. The results of two-way variance analysis of processes performance related to the selected industry in DIO are shown in Table 3.14.

The related formulas and calculations for the numerical application are as follows:

$$SS_{treatment} = \frac{1}{b}\sum_{i=1}^{a} Y_{io}^2 - \frac{\sum_{i=1}^{a} (Y_{io})^2}{a \times b} \tag{3.27}$$

In this formula, $SS_{treatment}$ is the sum of treatment data, Y_{io} is the sum total related to (*i*) row, (*a*) is the number of treatment states or rows, and (*b*) is the number of disorder states or columns. So, we have:

$$SS_{treatment} = \frac{1}{7}(792074.28) - \frac{(1258.35)^2}{14} = 50.27$$

Moreover, we have:

$$MS_{treatment} = \frac{SS_{treatment}}{a-1} \tag{3.28}$$

In this formula, $MS_{treatment}$ is the average sum related to the treatment data. Therefore, we have:

$$MS_{treatment} = \frac{50.27}{2-1} = 50.27$$

$$SS_{block} = \frac{1}{a}\sum_{j=1}^{b} Y_{oj}^2 - \frac{\sum_{j=1}^{b} (Y_{oj})^2}{a \times b} \tag{3.29}$$

TABLE 3.14

Two-Way Variance Analysis of Processes Performance in Selected Industry in DIO

Department name	Quality assurance	Material and product planning	Internal and external business	Financial and economics	Human resources	Health and safety	Warehouses management	Y_{io}
Spring & Summer	62.71	87.45	96.75	165	58.43	69.07	76.5	615.91
Autumn & Winter	106.28	94.64	119.75	81.75	42.31	96.71	101	642.44
Y_{oj}	168.99	182.09	216.50	246.75	100.74	164.78	177.50	1258.35

In this formula, SS_{block} is the sum of disorder data, Y_{oj} is the sum total related to (j) column, (a) is the number of treatment states or rows, and (b) is the number of disorder states or columns. So, we have:

$$SS_{block} = \frac{1}{2}(238610) - \frac{(1258.35)^2}{14} = 6201.81$$

Moreover, we have:

$$MS_{block} = \frac{SS_{block}}{b-1} \tag{3.30}$$

In this formula, MS_{block} is the average sum related to disorder data. Therefore, we have:

$$MS_{block} = \frac{6201.81}{7-1} = 1033.635$$

$$SS_{error} = \sum_{i=1}^{a}\sum_{j=1}^{b} Y_{ij}^2 - \frac{\left(\sum_{i=1}^{a}\sum_{j=1}^{b} Y_{ij}\right)^2}{a \times b} \tag{3.31}$$

In this formula, SS_{error} is the sum of error data, Y_{ij} is the sum total related to (i) row and (j) column, (a) is the number of treatment states or rows, and (b) is the number of disorder states or columns.

So, we have:

$$SS_{error} = 124824.84 - \frac{(1258.35)^2}{14} = 11718.65$$

Moreover, we have:

$$MS_{error} = \frac{SS_{error}}{(a-1)(b-1)} \tag{3.32}$$

In this formula, MS_{error} is the average sum related to the error data. Therefore, we have:

$$MS_{error} = \frac{11718.65}{(2-1)(7-1)} = 1953.11$$

$$\left\{ \begin{array}{l} F_{treatment} = \dfrac{MS_{treatment}}{MS_{error}} \hspace{4cm} (3.33) \\[1em] F_{block} = \dfrac{MS_{block}}{MS_{error}} \hspace{4.5cm} (3.34) \end{array} \right.$$

In these formulas, $F_{treatment}$ is the statistical distribution function for treatment and F_{block} is the statistical distribution function for disturbance. So, we have:

$$\begin{cases} F_{treatment} = \dfrac{50.27}{1953.11} = 0.026 \\[4mm] F_{block} = \dfrac{1033.635}{1953.11} = 0.53 \end{cases}$$

Now, if an interpretative framework is considered for the latter issue, we take a look at the following pattern:

$$\begin{cases} H_{.}: F_T \leq F_{\alpha, a-1, (a-1)(b-1)} \qquad \text{The treatment source does not make any significant difference} \\ H_1: F_T > F_{\alpha, a-1, (a-1)(b-1)} \qquad \text{The treatment source makes a significant difference} \end{cases}$$

(3.35)

$$\begin{cases} H_{.}: F_B \leq F_{\alpha, b-1, (a-1)(b-1)} \qquad \text{The disorder source does not make any significant difference} \\ H_1: F_B > F_{\alpha, b-1, (a-1)(b-1)} \qquad \text{The disorder source makes a significant difference} \end{cases}$$

(3.36)

Analysis:

$$F_{Time} = 0.026 , F_{0.05, 1, 6} = 5.99 \rightarrow 0.026 < 5.99$$

The **time period** does not make any significant difference

$$F_{Department} = 0.53 , F_{0.05, 6, 6} = 4.28 \rightarrow 0.53 < 4.28$$

The **kind of department** does not make any significant difference

3.5 IDENTIFICATION AND DETERMINATION OF THE DESIRED QET IN SUPPORT PROCESSES

These processes divided into relevant sub-processes based on the suggested QET are presented in Figure 3.15.

3.5.1 NUMERICAL APPLICATION OF QET IN THE DETERMINED SCOPE FOR SUPPORT PROCESSES

* HSE Management (*Failure Mode and Effects Analysis—FMEA*)

Health, Safety, and Environmental (HSE) management is one of the most important issues to be seriously considered by any organization in the 21st century. The best tool for carrying it out is through FMEA. But, before forming the main table, the

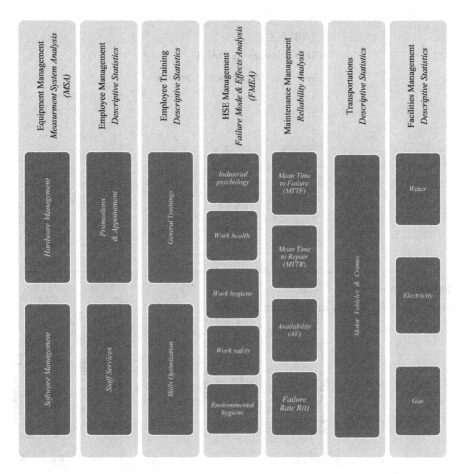

FIGURE 3.15   Desired QET in support processes.

triple issues must be considered as a prerequisite for the main table. Such a prerequisite table is presented in Table 3.15.:

The main table for FMEA is included in Table 3.16.

- Maintenance Management (*Reliability Analysis*)

The numerical application of the model in the reliability analysis section is related to the failure time of the Computer Numerical Control (CNC) machine linked up with the selected manufacturing factory of DIO. All failure times of the latter machine were gathered for 30 months. The data related to the failure time of the CNC machine for 30 months and the conclusions of Statistical Distribution Functions (SDF), together with relevant calculations for each are assembled in several sheets of an Excel program. Before doing this, the applicable and useful SDF are introduced:

TABLE 3.15

Triple issues in FMEA

Severity	Occurrence	Detection
1	1	10
2	2	9
3	3	8
4	4	7
5	5	6
6	6	5
7	7	4
8	8	3
9	9	2
10	10	1

TABLE 3.16

Main Table for FMEA in HSE management

N	Sub-process function	Potential failure mode	Potential effect(s) of failure	Severity	Potential cause (s) of failure	Occurrence	Current process control(s)	Detection	RPN = severity × occurrence × detection	Recommended action(s)	Responsibility and target completion date
1	Industrial psychology	Mental hazards	RD	8	RD	6	RD	8	384	RA	D
2	Work health	Health hazards	RD	7	RD	5	RD	5	175	RA	D
3	Work hygiene	Hygiene hazards	RD	6	RD	4	RD	5	120	RA	D
4	Work safety	Safety hazards	RD	6	RD	4	RD	4	96	RA	D
5	Environmental hygiene	Environmental hazards	RD	7	RD	5	RD	3	105	RA	D

RPN ≤ 150 → The sub-process is under control.
150 < RPN ≤ 250 → The sub-process must be contorlled.
RPN > 250 → The sub-process requires an action plan.
RD: Relevant descriptions, RA: Recommendation action(s), D: Date.

The popular statistical distribution functions usually deployed in the maintenance departments for calculating reliability are as follows:

3.5.1.1 Exponential Distribution Function

One of the best statistical distribution functions deployed for describing the life cycle of insulating oils, fluids (dielectrics), certain material, products, and the failure time of factories equipment is provided in Nelson, 2004. The probability density of this statistical distribution function is as follows:

$$f(t, \lambda) = \lambda e^{-\lambda t} \tag{3.37}$$

Where t is the failure time and λ is the failure rate. Regarding the definition of the reliability function, the following obtains:

$$R(t, \lambda) = \int_t^\infty f(t, \lambda) dt \tag{3.38}$$

So, we have:

$$R(t, \lambda) = [-e^{-\lambda t}]_t^\infty = e^{-\lambda t} \tag{3.39}$$

3.5.1.2 Ultra Exponential Distribution Function

This statistical distribution function is considered where equipment or systems with very short or very long breakdown times are utilized. Certain computers with breakdown times, very long at times, are suitable for applications in this type of distribution function (Nelson, 2004). The probability density of this statistical distribution function is as follows:

$$f(t, \lambda, k) = 2\lambda k^2 e^{-2\lambda k t} + 2\lambda(1-k)^2 e^{-2\lambda(1-k)t} \tag{3.40}$$

In this formula, t is the failure time and λ is the failure rate, and k is the constant value $0 \le k \le 0.5$. Regarding Equation (3.38) related to the definition of the reliability function, the following holds:

$$R(t, \lambda, k) = \left[-ke^{-2\lambda t} - (1-k)e^{-2\lambda(1-k)t} \right]_t^\infty = ke^{-\lambda t} + (1-k)e^{-2\lambda(1-k)t} \tag{3.41}$$

3.5.1.3 Gamma Distribution Function

This function, one of the traditional statistical distribution functions, is usually deployed for a large number of equipment or gears along with an accelerated test where the location parameter is a linear function of (possibly transformed) stress (Nelson, 2004). The probability density of this statistical distribution function is as follows:

$$f(t, \alpha, \beta) = \frac{1}{\beta^\alpha \Gamma(\alpha)} t^{\alpha-1} e^{-\frac{t}{\beta}} \tag{3.42}$$

α parameter is a unitless pure number. It is also designated as the "slope" parameter. β parameter is called the characteristic life. This parameter has the same units as t, e.g., hours, months, cycles, etc. The shape parameter (α) and the scale parameter (β) are positive.

3.5.1.4 Weibull Distribution Function

The Weibull distribution is often used for the product life since it models either increasing or simply decreasing failure rates. It is also used as a distribution for such product properties as strength (electrical or mechanical), resistance, etc., in accelerated tests. It is used to describe the life cycle of roller bearings, electronic components, ceramics, capacitors, and dielectrics in accelerated tests. According to the extreme value theory, it may describe a "weakest link" product. Such a product consists of many parts from the same life distribution, and the product fails with the first part of failure (Nelson, 2004). The probability density of this statistical distribution function is as follows:

$$f(t, \alpha, \beta) = \frac{\alpha}{\beta^{\alpha}} t^{\alpha-1} e^{-\left(\frac{t}{\beta}\right)^{\alpha}} \tag{3.43}$$

α parameter determines the shape of the distribution and β parameter determines the spread. And considering Equation (3.38) related to the definition of the reliability function, we have:

$$R(t, \alpha, \beta) = \left[e^{-\left(\frac{t}{\beta}\right)^{\alpha}} \right]_{t}^{\infty} = e^{-\left(\frac{t}{\beta}\right)^{\alpha}} \tag{3.44}$$

3.5.1.5 Normal Distribution Function

In a normal (or Gaussian) distribution function, the hazard function increases without limit. Thus, it may be used for describing products with wear-out failure. It has been used to describe the life span of incandescent lamp (light bulb) filaments and electrical insulations. It is also used as the distribution function for such product properties as strength (electrical or mechanical), elongation, and impact resistance in accelerated tests (Nelson, 2004). The probability density of this statistical distribution function is as follows:

$$f(t, \mu, \sigma) = \frac{1}{\sigma\sqrt{2\pi}} e^{-\left[\frac{(t-\mu)^2}{2\sigma^2}\right]} \tag{3.45}$$

In the above formula, μ is the population mean and may have any value, σ is the population standard deviation and must be positive, and μ and σ are in the same measurement units.

3.5.1.6 Log-Normal Distribution Function

The lognormal distribution is widely used for live data, including metal fatigue, solid-state components (semiconductors, diodes, etc.), and electrical insulations. The lognormal and normal distributions are related to each other, and this fact is used to analyze lognormal data with the same methods used on normal data (Nelson, 2004). The probability density of this statistical distribution function is as follows:

$$f(t, \mu, \sigma) = \frac{0.4343}{t\sigma\sqrt{2\pi}} e^{-\left[\frac{(\log(t)-\mu)^2}{2\sigma^2}\right]} \tag{3.46}$$

where μ is the mean of the log of life—not of life, μ is called the log mean and may have any value from $-\infty$ to $+\infty$, σ is the standard deviation of the log of life—not of life, σ is called the log standard deviation and must be positive, and μ and σ are not "times" like t, rather, they are unitless pure numbers. Moreover, the relevant formulas which can be used are as follows (Nelson, 2004) (Porter, 2004):

$$\text{Total Number of Failures} = \sum_{i=1}^{j} \text{NF}_i \quad (3.47)$$

$$\text{Failure Probability (FP}_i) = \frac{\text{NF}_i}{\sum_{i=1}^{j} \text{NF}_i} \quad (3.48)$$

$$\text{Total Failure Probability} = \sum_{i=1}^{j} \text{FP}_i \quad (3.49)$$

$$\text{Total AT}_i \times \text{FP}_i \text{ (Average Time of Failure)} = \sum_{i=1}^{j} \text{AT}_i \times \text{FP}_i \quad (3.50)$$

$$\text{Number of Failure Average} = \frac{1}{\sum_{i=1}^{j} \text{AT}_i \times \text{FP}_i} \quad (3.51)$$

$$\text{Variance} = \frac{1}{j-1}\left[\sum_{i=1}^{j}(\text{FT}_i)^2 - \frac{1}{j}\sum_{i=1}^{j}(\text{FT}_i)^2\right] \quad (3.52)$$

$$\sigma = \sqrt{\frac{1}{j-1}\left[\sum_{i=1}^{j}(\text{FT}_i)^2 - \frac{1}{j}\sum_{i=1}^{j}(\text{FT}_i)^2\right]} \quad (3.53)$$

$$\chi^2 = \sum_{i=1}^{j}\frac{(P(i)-\text{FP}_i)^2}{\text{FP}_i} \quad (3.54)$$

After finding the proper statistical distribution function (in this application, the exponential distribution function), we can refer to the related formula for calculating the reliability of the mentioned equipment or machine (in this application, a CNC machine). Therefore, we have the formula $R(t,\lambda) = [-e^{-\lambda t}]_t^{\infty} = e^{-\lambda t}$

Then we will have: $R(t,\lambda) = e^{-\lambda t} \Rightarrow R(100, 0.0197) = e^{-1.97} = 0.1394 \approx 0.14 \Rightarrow 14\%$

All this means that considering the failure times of this CNC machine, the probability with which this CNC machine can work without any failure or breakdown (e.g., during 100 hours) is 14%.

Figure 3.16 presents the information in this regard (Karbasian and Rostamkhani, 2019).

Failure times data for a CNC machine in 30 months							
N	Failure Time (hour)	Time Range (hour)	Average Time (AT)	Fr	Number of Failures (NF)	Failure Probability (FP)	AT×FP
1	1.25	0-24.99	12.5	0	20	0.67	8.33
2	2.35						
3	3.75	25-49.99	37.5	25	2	0.07	2.50
4	5.50						
5	4.55	50-74.99	62.5	50	1	0.03	2.08
6	7.85						
7	6.50	75-99.99	87.5	75	1	0.03	2.92
8	8.85						
9	6.45	100-124.99	112.5	100	1	0.03	3.75
10	7.50						
11	10.25	125-149.99	137.5	125	1	0.03	4.58
12	12.65						
13	14.35	150-174.99	162.5	150	1	0.03	5.42
14	16.25						
15	18.00	175-199.99	187.5	175	1	0.03	6.25
16	19.95						
17	21.00	200-224.99	212.5	200	1	0.03	7.08
18	22.85						
19	23.95	225-249.99	237.5	225	1	0.03	7.92
20	24.15						
21	32.75						
22	36.05						
23	51.12						
24	77.22						
25	120.00						
26	138.50						
27	155.55						
28	177.75						
29	210.25						
30	228.00						
	Total				30	1.00	50.83
Var	4444.3167	σ	66.6657		M : Average Time of Failure		50.83
	Number of Failure Average =1/M						0.0197
Statistical Distribution Functions							

Failure Time

250.00
200.00
150.00
100.00
50.00
0.00
    0   10   20   30   40

Number of Failures

20
2 1 1 1 1 1 1 1 1 1
1 3 5 7 9 11 13 15 17 19

N	Title of Statistical Distribution Function	Chi-square(χ^2)	Chi-square ($\chi^2_{0.001,8}$)	Result
1	Exponential Distribution Function	27.8646		OK
2	Ultra Exponential Distribution Function	28.5759		NOT OK
3	Gamma Function	30.1277		NOT OK
4	Weibull Function	30.0035	27.8680	NOT OK
5	Normal Function	29.7544		NOT OK
6	Log-normal Function	28.4382		NOT OK

FIGURE 3.16 Failure times data for a CNC machine.

Additional explanations and mathematical calculations results for choosing the best Statistical Distribution Function (SDF) have been attached to Appendix.

EXERCISES

3.1. Identify the main components of Supply Chain Management (SCM) and form the required assessment matrices between them and Quality Engineering Techniques (QET). How can this model be realized (as a project)?

3.2. Identify the main domains of Industrial Engineering (IE) and form the required assessment matrices between them and Quality Engineering Techniques (QET). How can this model be realized (as a project)?

4 Results of Implementing the Model

4.1 INTRODUCTION

This chapter presents the results of implementing the model. The most important results attained through executing the proposed model in industrial organizations are as follows:

- Augmenting of productivity and sustainability
- Creation of added values in organizational processes
- Identification of risks and opportunities in organizational processes
- Knowledge of the method to assess the components of industrial economics via statistical and non-statistical means of the quality engineering techniques (QET) model.

4.2 Z-MR CONTROL CHARTS FOR DESCRIBING THE IMPACTS OF DIFFERENT UNITS ON PRODUCTIVITY AND SUSTAINABILITY

The Process Capability Indices (PCI) can be defined for a selected QET—for example, in this book, the Z-MR control charts are considered to be a strong analytical and graphical tool—where the clustered bar chart for each unit of the selected industry and the relevant details are displayed in Figures 4.1a to 4.1g. It should be noted that these operational tables describe the integrated objectives in the form of whole operations that each section must accomplish. In other words, these applied tables contain more details where the sub-processes and related indices are introduced. One of the most important and innovative aspects incorporated in the current research, for the first time, is that the relevant data (in this book, the mean value) of all subunits are controlled by Z-MR control charts (statistical process control) for the main data (in this book, the target value). It should be remembered that under these conditions, the benefits of relevant indices are underscored. Hence, the first chart in all of Figures 4.1a to 4.1g includes the Z-MR control chart and the second chart in all Figures 4.1a to 4.1g includes relevant C_{pmk} for these indices. In each figure, the unit condition, the number of sub-objectives, the related sub-processes, and other descriptions are explained at the top of each section. What we intend to show is that the proposed model functions and covers the productivity and sustainability concepts leading up to added values in organizational processes through related calculations. It is to be remembered that the

Research & Development Unit

There are four sub-objectives in this unit located in one sub-process designated as product design. The Mean Value (μ) data are under control and compared with the Target Value (T) and the related C_{pmk} for the four indices of one sub-process can easily be calculated. The performance of this unit can have a direct influence on **sustainability** in the organization.

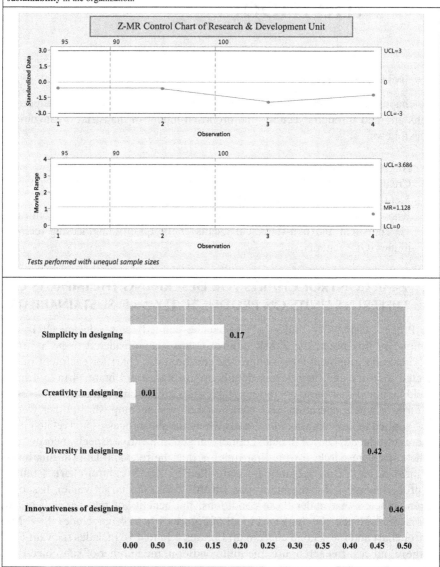

FIGURE 4.1A   Process capability indices in Research & Development Unit.

Quality Assurance Unit

There are seven sub-objectives in this unit located in nine sub-processes designated as follows: Input quality - Output quality - Customer relationship - Customer satisfaction - Sale services - Production control - Production productivity - Production sustainability - Establishment of IMS. The Mean Value (μ) data are under control and compared to the Target Value (T) and the related C_{pmk} for 18 indices of these nine sub-processes can easily be calculated. The performance of this unit can influence both **productivity** and **sustainability**.

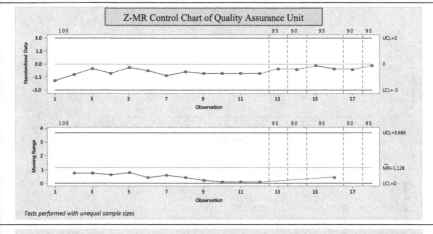

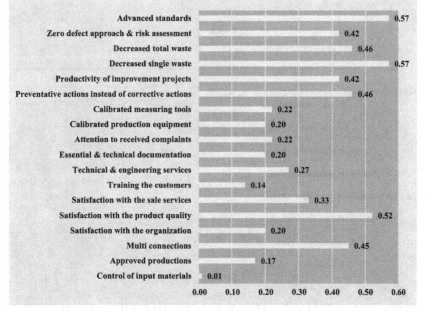

FIGURE 4.1B Process capability indices in Quality Assurance Unit.

Planning & Schedule Unit

There are two sub-objectives in this unit located in five sub-processes designated as follows: Control and orders management - Production planning - Hardware management - Software management - Production productivity. The Mean Value (μ) data are under control and compared to the Target Value (T) and the related C_{pmk} for nine indices of these five sub-processes can easily be calculated. The performance of this unit can influence **productivity**.

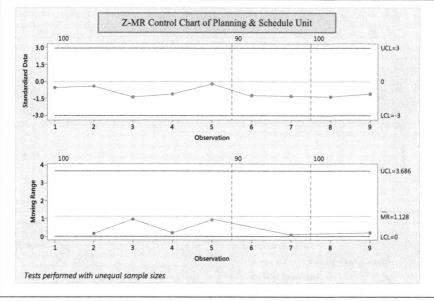

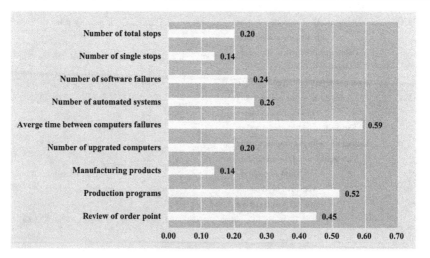

FIGURE 4.1C   Process capability indices in Planning & Schedule Unit.

Trade & Commercial Unit

There are three sub-objectives in this unit located in seven sub-processes designated as follows: Orders reception - Products delivery - Supplier selection - Supplier assessment - Market development - Lateral items sales - Export. The Mean Value (μ) data are under control and compared to the Target Value (T) and the related C_{pmk} for seven indices of these seven sub-processes can easily be calculated. The performance of this unit can influence **productivity**.

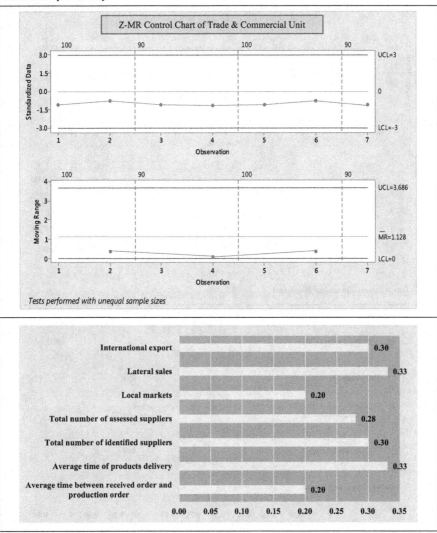

FIGURE 4.1D   Process capability indices in Trade & Commercial Unit.

Human Resources Unit

There are three sub-objectives in this unit located in five sub-processes designated as follows: Promotion and appointment - Staff services - General training - Skills optimization - Knowledge optimization. The Mean Value (μ) data are under control and compared to the Target Value (T) and the related C_{pmk} for five indices of these five sub-processes can easily be calculated. The performance of this unit can influence **productivity**.

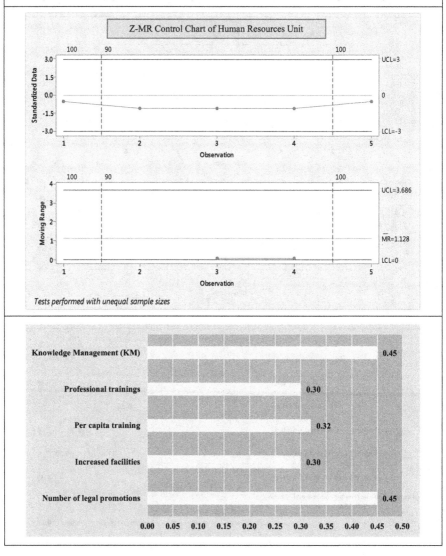

FIGURE 4.1E Process capability indices in Human Resources Unit.

Health & Safety & Environment Unit (HSE)

There are nine sub-objectives in this unit located in five sub-processes designated as follows: Industrial psychology - Work health - Work hygiene - Work safety - Environmental hygiene. The Mean Value (μ) data are under control and compared to the Target Value (T) and the related C_{pmk} for nine indices of these five sub-processes can easily be calculated. The performance of this unit can influence both **productivity** and **sustainability**.

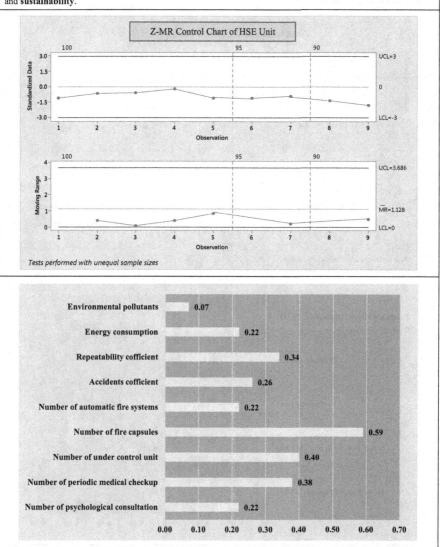

FIGURE 4.1F   Process capability indices in Health & Safety & Environment (HSE) Unit.

Maintenance Unit

There are six sub-objectives in this unit located in five sub-processes designated as follows: Mean time to failure (MTTF) - Mean time to repair (MTTR) - Availability (AV) - Failure rate R (T) - Total productive maintenance (TPM). The Mean Value (μ) data are under control and compared to the Target Value (T) and the related C_{pmk} for five indices of these five sub-procetsses can easily be calculated. The performance of this unit can influence both **productivity** and **sustainability**.

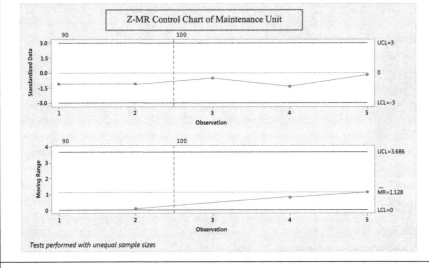

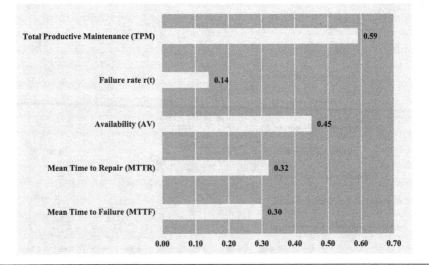

FIGURE 4.1G   Process capability indices in Maintenance Unit.

TABLE 4.1
Role of Impact of Each Unit on Productivity and Sustainability

	Impacts	
Units	Productivity	Sustainability
Research & Development		*
Quality Assurance	*	*
Planning & Scheduling	*	
Trade & Commercial	*	
Human Resources	*	
HSE	*	*
Maintenance	*	*

relevant calculations for C_{pm} and C_{pmk} are defined based on the following formulas (Karimi Gavareshki et al., 2019):

$$C_{pm} = \frac{USL - LSL}{6\sqrt{\sigma^2 + (\mu - T)^2}}, C_{pmk} = Min\left[\frac{USL - \mu}{3\sqrt{\sigma^2 + (\mu - T)^2}}, \frac{\mu - LSL}{3\sqrt{\sigma^2 + (\mu - T)^2}}\right] \quad (4.1)$$

where USL is the upper specification limit and LSL is the lower specification limit. σ^2 is the variance, μ is the mean value, and T is the target value. It should be pointed out that in all calculations, the standard deviation is $\sigma = 10$ by default, the tolerance range in the field of indices is %100 ± %20, the target value (T) in all processes is not equal to the mean value (μ), and the midpoint of technical specification limits is named (M). Generally speaking, the process capability indices are compared; in fact, a couple of indices (C_p, C_{pk}) are compared with their respective (C_{pm}, C_{pmk}). The impacts of different units on productivity and sustainability are presented in Table 4.1 (Karimi Gavareshki et al., 2019).

4.3 TOTAL SCORE CALCULATION OF PRODUCTIVITY AND SUSTAINABILITY BEFORE AND AFTER IMPLEMENTING QET

The total scores for productivity and sustainability—before implementing the selected QET—are usually figured out by means of simple numerical methods found in any books on elementary statistics. However, since there are applications related to different statistical and non-statistical techniques in QET, in order to find out the total score of productivity and sustainability after implementing the selected QET, we preferred to use Analytic Hierarchy Process (AHP) analysis. In this book, we found the latter score in 6 values (between 1 and 9). The relevant calculations were conducted considering the general conditions of the selected industry. These calculations, together with more details, are provided in Table 4.2.

TABLE 4.2

AHP Analysis for Statistical and Non-Statistical Techniques in Productivity and Sustainability

AHP analysis for statistical and non-statistical techniques in productivity and sustainability

Standard statistical tools based on ISO10017	Average	I_1(Index)	Sub-index		T_1	
Descriptive statistics	5.58	0.83	I_{111}	0.08	T_{111}	0.46
Design of experiments (DOE)	5.92		I_{112}	0.09	T_{112}	0.52
Statistical process control (SPC)	7.25		I_{113}	0.11	T_{113}	0.78
Statistical hypothesis test	6.58		I_{114}	0.10	T_{114}	0.64
Process capability analysis	5.42		I_{115}	0.08	T_{115}	0.43
Statistical tolerances	4.75		I_{116}	0.07	T_{116}	0.33
Time series analysis	6.67		I_{117}	0.10	T_{117}	0.66
Regression analysis	6.25		I_{118}	0.09	T_{118}	0.58
Reliability analysis	7.25		I_{119}	0.11	T_{119}	0.78
Simulation	5.15		I_{120}	0.08	T_{120}	0.39
Sampling	6.82		I_{121}	0.10	T_{121}	0.69
Total average	6.15		Sum	1.00	Sum	6.26
Non-statistical tools	**Average**	**I_2(Index)**	**Sub-index**		**T_2**	
Quality function deployment (QFD)	5.00	0.17	I_{21}	0.35	T_{21}	1.74
Value engineering (VE)	4.75		I_{22}	0.33	T_{22}	1.57
Value stream mapping (VSM)	4.58		I_{23}	0.32	T_{23}	1.46
Workflow analysis (WFA)	4.12		I_{24}	0.29	T_{26}	1.18
Cost of quality (COQ)	3.98		I_{25}	0.28	T_{27}	1.11
Failure mode effects analysis (FMEA)	6.15		I_{26}	0.43	T_{28}	2.64
Designing failure mode effects analysis (DFMEA)	5.75		I_{27}	0.40	T_{29}	2.31
Production failure mode effects analysis (PFMEA)	4.65		I_{28}	0.32	T_{30}	1.51
Total average	4.87	1.00	Sum	1.00	Sum	4.78
Sum				$T = (I_1 \times T_1) + (I_2 \times T_2)$		
T= Total score of productivity and sustainability (between 1 and 9)				**6.00**		

It can be seen that, the total score for productivity and sustainability after implementing the selected QET equals 6.00.

The following relevant formulas can be used in this regard (Karimi Gavareshki et al., 2018):

$$I_1 = \frac{\sum_{i=1}^{m} A_i}{\sum_{i=1}^{m} A_i + \sum_{j=1}^{n} A_j}, \; I_2 = \frac{\sum_{j=1}^{n} A_j}{\sum_{i=1}^{m} A_i + \sum_{j=1}^{n} A_j} \qquad (4.2)$$

where A_i is the average of the statistical techniques and A_j is the average of the non-statistical techniques. I_1 is the main index of the statistical techniques and I_2 is the main index of the non-statistical techniques.

$$I_{1k} = \frac{A_i}{\sum\limits_{i=1}^{m} A_i}, \quad k = 11,\ldots,21, \; I_{2l} = \frac{A_j}{\sum\limits_{j=1}^{n} A_j}, \quad l = 1,2,3 \tag{4.3}$$

I_{1k} is the sub-index of the statistical techniques and I_{2l} is the sub-index of the non-statistical techniques.

$$T_{1k} = A_i \times I_{1k}, \quad k = 11,12,\ldots, \quad i = 1,\ldots,m \tag{4.4}$$

$$T_{2l} = A_j \times I_{2l}, \quad l = 1,2,\ldots, \quad j = 1,\ldots,n \tag{4.5}$$

$$T_1 = \sum_{k=11} T_{1k}, \; T_2 = \sum_{l=1} T_{2l} \tag{4.6}$$

T_{1k} is the multiple factors of the statistical techniques and T_{2l} is the multiple factors of the non-statistical techniques. But, T_1 is the total multiple factors of the statistical techniques and T_2 is the total multiple factors of the non-statistical techniques.

$$I_1 + I_2 = 1, \; \sum_{k=11} I_{1k} = 1, \; \sum_{l=1} I_{1l} = 1 \tag{4.7}$$

In order to calculate the total score of productivity and sustainability (T), the following formula can be used:

$$T = (I_1 \times T_1) + (I_2 \times T_2) \tag{4.8}$$

4.4 DEMONSTRATION OF THE GROWTH IN ADDED VALUES FOR INDUSTRIAL PRODUCTS MANUFACTURING

It should be obvious that the classification of organizational processes and sub-processes and extracting their respective C_{pmk} can help determine the action plans to increase the processes capability **eventually leading up to added values in triple organizational processes.** Note that the impact of applying the selected QET (in the form of an action plan) can result in added values for industrial products manufacturing. A schematic representation is provided in Figure 4.2. Also, a sample action plan with details is provided in Table 4.3.

FIGURE 4.2   Creating added values by applying QET.

TABLE 4.3

Action Plan for Creating Added Values in Some Organizational Processes

Process	C_{pmk} (before)	Plan details	C_{pmk} (after)
Orders	0.32	• Applying related techniques of statistical techniques (ST) at all levels of reception e.g., descriptive statistics	0.52
Planning	0.29	• Applying related techniques of quality engineering techniques including statistical techniques (ST) such as statistical tolerances, simulation, and analysis of time series; and such non-statistical techniques (NT) as QFD, VE, and DFMEA	0.49
Manufacturing	0.20	• Applying related techniques of statistical techniques (ST) at all levels of production e.g., descriptive statistics and PFMEA	0.28
Quality Control	0.17	• Calibrating all tools related to quality control • Applying such related techniques of statistical techniques (ST) as SPC at all levels of control • Applying such related techniques of non-statistical techniques (NT) as COQ at all levels of control • Considering risks and opportunities in control inputs and outputs	0.25
Customer	0.32	• Anticipating customer requirements • Applying related techniques of SPC in assessing customer satisfaction • Considering risks and opportunities in customer requirements	0.45
Productivity Management	0.33	• Manufacturing resources management • Planning and manufacturing control • Manufacturing timing management • Quality system management • Projects control management	0.52
Sustainability Management	0.33	• Manufacturing processes design • Manufacturing risk management • Product reliability management • Production waste management	0.48

TABLE 4.4
Survey Results in Generalizing Proposed Model

Effective factors in generalizing proposed model	Average (ST)	Average (N-ST)
Comprehensiveness	7.55	6.35
Feasibility	8.15	7.45
Flexibility	6.12	5.78
Stability	7.82	6.66
Power	6.75	5.55
Total average (between 1 and 9)	7.28	6.36
Percentage of total averages	80.87	70.64
Success in generalization (between 1 and 9)	7.10	
Percentage of success	78.92	

ST: Statistical techniques, N-ST: Non-Statistical techniques.

TABLE 4.5
Research Data Reliability

Reliability statistics		
Cronbach's alpha	Cronbach's alpha based on standardized items	Number of items
0.82	0.825	5

4.5 GENERALIZING THE ASSESSMENT RESULTS OF THE MODEL TO OTHER ORGANIZATIONS

In order to generalize the assessment results of the model, our questionnaires were administered among 26 representative managers of some industries affiliated with the Defense Industries Organization (DIO). The results of this survey are presented in Table 4.4 (see Appendix).

The research reliability values for the 26 management representatives of some industries affiliated with the DIO are presented in Table 4.5.

4.6 ASSESSING THE MODEL'S RISKS AND OPPORTUNITIES IN MANUFACTURING INDUSTRIES

Risks and opportunities management is a continuous process and, if properly implemented, can help keep all constituents of organization in a continuously improving state. Risks and opportunities management is illustrated in Figure 4.3.

FIGURE 4.3   Risks and opportunities management process.

4.6.1   RISKS AND OPPORTUNITIES IDENTIFICATION

By reviewing a source list of all potential risks and opportunities along with past experiences, the organizational risks and opportunities are identified in all three process categories: main, leadership, and support. The role of risks and opportunities in relation to all three categories of processes is indicated in Figure 4.4.

The following are the risks identified in the main processes (Figure 4.5).

The following are the risks identified the leadership processes (Figure 4.6).

The following are the risks identified in the support processes (Figure 4.7).

4.6.2   RISKS AND OPPORTUNITIES ASSESSMENT

By using measurement tools, the risks and opportunities should be categorized and prioritized. The number of risks and opportunities usually exceeds the organizational capacity to deploy a specific system to analyze them and design emergency plans. The prioritization process helps organizations to prioritize the risks and opportunities that are more urgent and more likely to occur. It is recommended that risks and opportunities be prioritized by using the data given in Table 3.15.

4.6.3   RESPONDING TO RISKS AND OPPORTUNITIES

The classic approach for dealing with risks and opportunities is to move from identifying the issue to its depth. However, it must first be determined how risks and

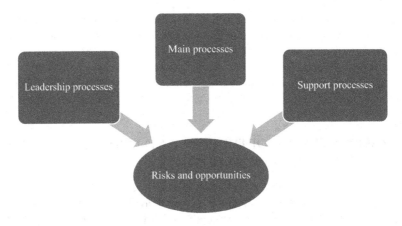

FIGURE 4.4   Relationship between risks and opportunities and three categorized processes.

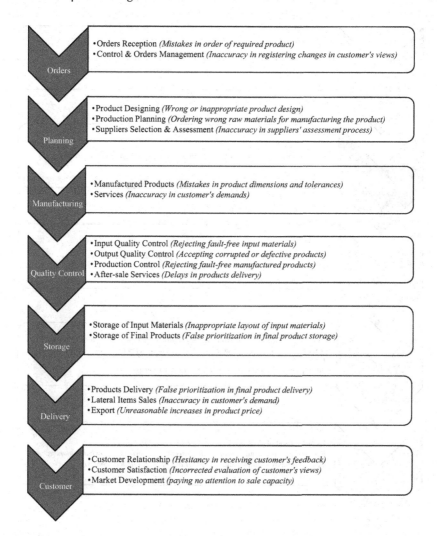

FIGURE 4.5   Identified risks in main processes.

opportunities can be better managed and the root causes of identified risks and opportunities be recognized. In this regard, the following questions might be asked:

- What are the causes of these risks and opportunities?
- How can these risks and opportunities affect the organization?

Generally, in responding to risks, the following occurs:

- Avoidance: Eliminating a particular risk or threat by removing its cause (preventive action).
- Adjustment: Reducing the expected financial value of a particular risk by reducing its occurrence.

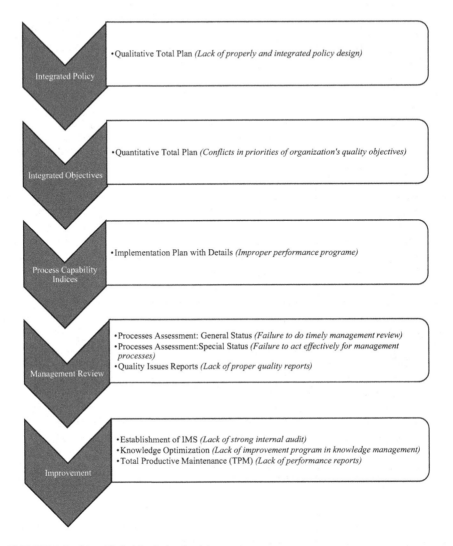

FIGURE 4.6 Identified risks in leadership processes.

- Acceptance: Accepting the consequences of the risk, which most often occurs by formulating and implementing an emergency plan for an event that is likely to happen. This emergency plan can be implemented for a short term and will last in the organization for as long as the risk threat continues.

Responding to opportunities generally includes:
- Modification: Exploiting the advantages of the created positive outcomes of that opportunity on a permanent or temporary basis.

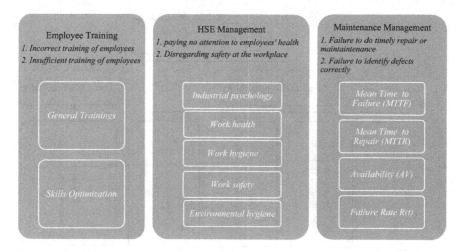

FIGURE 4.7   Identified risks in support processes.

4.7   IMPACTS OF IMPLEMENTING THE MODEL IN INDUSTRIAL ECONOMICS

The industrial economics is one of the most important issues studied since the beginning of the 21st century. In a valuable research study, this issue is divided into two categories (Federico, 2016):

- Empirical economics
  This includes estimating demand functions as well as cost functions, measuring market power and power assessment of market entrance. It can also include a variety of activities focused on the impact of particular strategies in competitive markets.
- Theoretical economics
  This includes assessing different models under different circumstances of application. Generally, this branch of economics deals with simulation processes.

The best quality engineering techniques for both branches of economics are presented in Figure 4.8.

The proposed quality engineering techniques which contain both statistical and non-statistical procedures can help us achieve a better understanding of relevant components in industrial economics. As a matter of fact, these techniques can provide robust tools for decision makers in implementing any short- or long-term plans in economics.

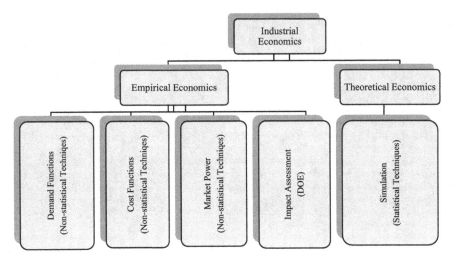

FIGURE 4.8   Suggested QET for industrial economics.

EXERCISE

4.1. The results related to the importance and performance of implementing the
QET model in a quality management system (QMS), in one industry, are
displayed in the table below. (a Likert scale is used from 1 to 9.)

N	Techniques	Average of importance	Variance of importance	Average of performance	Variance of performance
A	Descriptive statistics	6.83	3.24	3.17	2.5
B	Design and analysis of experiments	7.17	2.5	3	2.9
C	Statistical hypothesis tests	5	4.36	2.83	1.8
D	Process capability analysis	7	3.64	3	2.18
E	Regression analysis	3.83	6.15	2	1.1
F	Reliability analysis	3.67	4.6	2.5	2.27
G	Sampling	6.83	4	3.17	1.8
H	Simulation	4.17	6.88	2	1.1
I	Statistical process control charts	6.17	3.24	2.83	2.5
J	Statistical tolerances	4.17	4.7	2.33	2.42
K	Time series analysis	5	3.64	2.67	2.06

How can the above table, in both sections including importance and performance,
be analyzed?

5 Quality Engineering Techniques from Past to Future

5.1 A DISCUSSION ON COMPARISON OF PREVIOUS RESEARCH WITH THE PROPOSED MODEL

Of the previous research done on statistical techniques (ST) and non-statistical techniques (N-ST), the research that focused on singular applications in several industries or scientific fields are valuable. Table 5.1, for instance, sheds light on the real path of the research conducted so far.

Not one of the previous research studies has a comprehensive model for implementing QET, in particular, for industrial factories or manufacturing industrial products. Moreover, the process approach for describing the proposed model is the key point of the book.

Figure 5.1 presents the two main approaches for applying statistical and non-statistical techniques from the past to the present.

TABLE 5.1
Path of Research on (ST) and (N-ST) Applications

N	Author(s)	Research title
1	(Sima et al., 2019)	Feasibility of Using Simulation Technique for Line Balancing in Apparel Industry
2	(Tulcidas et al., 2019)	Statistical methodology for scale-up of an anti-solvent crystallization in the pharmaceutical industry
3	(Barua et al., 2018)	Chapter 10 – Statistical Techniques in Pharmaceutical Product Development
4	(Andrade et al., 2018)	Application of waves trapping statistical technique to estimate an extreme value in train aerodynamics
5	(Cristovao et al., 2018)	Fish canning industry wastewater variability assessment using multivariate statistical methods
6	(Memon and Shaikh, 2016)	Confidence bounds for energy conservation in electric motors: An economical solution using ST
7	(Lim and Antony, 2016)	Statistical process control readiness in the food industry: Development of a self-assessment tool
8	(Lin et al., 2015)	Identifying water recycling strategy using multivariate statistical analysis for high-tech industries

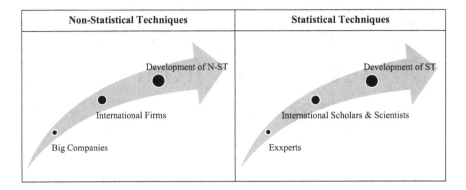

FIGURE 5.1   Trend of two main approaches in the development of QET.

A mixture of both ST and N-ST approaches in the form of an integrated model for industrial organizations has not been studied or at least has not been reported. For the first time, such an approach has been considered and applied in this research. Furthermore, several aspects of the proposed model in this book offer new approaches for carrying out systematic studies in industrial organizations.

5.2   FINAL RESULTS OF THE RESEARCH IN MANUFACTURING INDUSTRIES

Figure 5.2 provides a summary of the main points of the proposed model in this research.

The most creative aspect of this book is the introduction of a process map within the organization, besides exploiting several quality engineering techniques including statistical and non-statistical tools applied to different levels of organization. The most innovative recommendation of this book is to execute the proposed model in an

Leadership processes		
Number of Processes	Number of Sub-processes	Number of Techniques
5	9	6

Main processes		
Number of Processes	Number of Sub-processes	Number of Techniques
7	19	16

Support processes		
Number of Processes	Number of Sub-processes	Number of Techniques
7	19	4

FIGURE 5.2   Proposed model of QET for industrial factories.

effective format for the defense sectors of the country and to create or increase added values for manufacturing industrial products. The proposed model in this book utilizes several features as follows:

- Z-MR control charts for describing the impacts of units on productivity and sustainability
- Calculations of the total score of productivity and sustainability before and after implementing the model
- Demonstration of the growth in added values for manufacturing industrial products
- Generalization of the assessment results of the model to other organizations
- Assessment of risks and opportunities with using the model in manufacturing industries
- Impacts of implementing the model in industrial economics

5.3   SUGGESTIONS FOR FUTURE RESEARCH

One of the best suggestions that can be made is to design and implement the proposed model in fuzzy environments in industrial organizations. The fuzzy environments require advanced statistical and non-statistical techniques. Moreover, analytical mathematics must be considered. Furthermore, the uncertainty issue related to different results in statistical calculations is one of the most important points to consider and it seems to be the greatest challenge of the 21st century. The generalization of the proposed model to uncertain situations can be regarded as an attractive project which could change the future of industrial organizations. In the end, it is hoped that the proposed approach can lead to productivity and sustainability through the application of applying QE. The authors hope that the model proposed in this book could open a new door toward achieving superior goals in the industrial world of today.

References

Abbasi, M., Rostamkhani, R., (2014), Reliability Application of Industrial Manufacturing Networks in Outsourcing, *Journal of Engineering and Quality Management*, Volume 3, No. 4, Pages 247–259.

Andrade, A.R., Johnson, T., Stow, J., (2018), Application of Waves Trapping Statistical Technique to Estimate an Extreme Value in Train Aerodynamics, *Journal of Wind Engineering and Industrial Aerodynamics*, Volume 175, Pages 419–427. https://doi.org/10.1016/j.jweia.2018.02.009

Barua, A., Deb, P.K., Maheshwari, R., Tekade, R.K., (2018), Chapter 10—Statistical Techniques in Pharmaceutical Product Development, In R.K. Tekade (ed.), *Dosage Form Design Parameters*, Volume 2, Pages 339–362, National Institute of Pharmaceutical Education and Research (NIPER), Ahmedabad, India. https://doi.org/10.1016/B978-0-12-814421-3.00010-5

Bounazef, D., Chabani, C., Idir, A., Bounazef, M., (2014), Management Analysis of Industrial Production Losses by the Design of Experiments, Statistical Process Control, and Capability Indices, *Journal of Business and Management*, Volume 2, No. 1, Pages 65–72. https://doi.org/10.4236/ojbm.2014.21009

Cristovao, R.O., Pinto, V.M.S., Goncalvez, A., Martins, R.J.E., Loureiro, J.M., Boaventura, R.A.R., (2018), Fish Canning Industry Wastewater Variability Assessment using Multivariate Statistical Methods, *Process Safety and Environmental Protection*, Volume 102, Pages 263–276. https://doi.org/10.1016/j.psep.2016.03.016

Domingues, P., Sampaio, P., Arezes, P.M., (2016), Integrated Management Systems Assessment: A Maturity Model Proposal, *Journal of Cleaner Production*, Volume 124, No. 15, Pages 164–174. https://doi.org/10.1016/j.jclepro.2016.02.103

Espinosa-Garza, G., Loera-Hernandez, I., Antonyan, N., (2017), Increase of Productivity through the Study of Work Activities in the Construction Sector, *Manufacturing Engineering Society International Conference, 28–30 June, Vigo (Pontevedra)— Spain*, Pages 1003–1010. https://doi.org/10.1016/j.promfg.2017.09.100

Federico, E., (2016), Research in Economics and Industrial Organization, *Journal of Research in Economics*, Volume 70, No. 4, Pages 511–517. https://doi.org/10.1016/j.rie.2016.10.002

Fisher, C., (2014), New Techniques in Project Management, *American Journal of Industrial and Business Management*, Volume 4, No. 12, Pages 739–750. https://doi.org/ 10.4236/ajibm.2014.412080

Karbasian, M., Rostamkhani, R., (2017), Application of Design of Experiments Technique in Quality Management System for Assessing Various Factors Affecting Processes Performance of Defense Industries Organization, *Sharif Journal of Science and Technology*, Volume 33.1, No. 1.1, Pages 95–102. https://doi.org/10.24200/J65.2017.5579

Karbasian, M., Rostamkhani, R., (2019), Achieving Productive Reliability through Applying Statistical Distribution Functions, *International Journal of Quality and Reliability Management*. https://doi.org/10.1108/IJQRM-11-2018-0298

Karimi Gavareshki, M.H., Abbasi, M., Karbasian, M., Rostamkhani, R., (2018), Application of Quality Engineering Techniques in the Main Domains of Industrial Engineering, *Journal of Achievements in Materials & Manufacturing Engineering*, Volume 1, No. 90, Pages 22–40. https://doi.org/10.5604/01.3001.0012.7972

Karimi Gavareshki, M.H., Abbasi, M., Karbasian, M., Rostamkhani, R., (2019), Presenting a Productive and Sustainable Model of Integrated Management System for Achieving an

Added Value in Organisational Processes, *International Journal of Productivity and Quality Management.* https://doi.org/10.1504/IJPQM.2019.10023794

Karimi Gavareshki, M.H., Abbasi, M., Rostamkhani, R., (2017), Application of QFD and VE and Lean Approach for Control Tests in a Product Design, *Archives of Materials Science and Engineering*, Volume 84, No. 2, Pages 65–78. https://doi.org/10.5604/01.3001.0010.0980

Karimi Gavareshki, M.H., Sharifi Zamani, M., Rostamkhani, R., (2014), Identification and Determination of Effective Application of Statistical Prioritize Techniques in Quality Management System in Defense Industries Organization, *Iranian Electric Industry Journal of Quality and Productivity*, Volume 2, No. 4, Pages 18–29.

Lim, A.H.S., Antony, J., (2016), Statistical Process Control Readiness in the Food Industry: Development of a Self-Assessment Tool, *Trends in Food Science and Technology*, Volume 58, Pages 133–139.

Lin, S.W., Lee, M., Huang, Y.C., Den, W., (2015), Identifying Water Recycling Strategy using Multivariate Statistical Analysis for High-Tech Industries in Taiwan, *Resources Conservation and Recycling*, Volume 94, Pages 35–42.

Memon, A.J., Shaikh, M.M., (2016), Confidence Bounds for Energy Conservation in Electric Motors: An Economical Solution using Statistical Techniques, *Energy*, Volume 109, Pages 592–601. https://doi.org/10.1016/j.energy.2016.05.014

Montgomery, D., (1996a), *Statistical Quality Control* (Eighth Edition), translated to Farsi by Noorossana, R., Iran University of Science and Technology, Tehran, Iran.

Montgomery, D., (1996b), *Design and Analysis of Experiments* (First Edition), translated to Farsi by Noorossana, R., Iran University of Science and Technology, Tehran, Iran.

Nelson, W., (2004), *Accelerated Testing: Statistical Models, Test Plans and Data Analyses*, John Wiley & Sons, Hoboken, New Jersey, USA.

Olawumi, T.O., Chan, D.W.M., (2018), A Scientometric Review of Global Research on Sustainability and Sustainable Development, *Journal of Cleaner Production*, Volume 183, Pages 231–250. https://doi.org/10.1016/j.jclepro.2018.02.162

Plag, I., (2006), *Encyclopedia of Language & Linguistics* (Second Edition), University of Siegen, Siegen, Germany, Pages 121–128.

Porter, A., (2004), *Accelerated Testing and Validation*, Elsevier, Burlington, MA, USA, Page 01803.

Rajnoha, R., Sujova, A., Dobrovic, J., (2012), Management and Economics of Business Processes Added Value, *Procedia, Social and behavioral Sciences*, Volume 62, Pages 1292–1296. https://doi.org/10.1016/j.sbspro.2012.09.221

Rezaei, K., (2001), Using of the Quality Engineering Techniques in the Framework of Quality Management Systems, *Second International Quality Management Conference, CQM02 - 002, 22–25 July, Tehran—Iran.*

Rezazadeh, S., Jahani, A., Makhdoum, M., Meigooni, H.G., (2017), Evaluation of the Strategic Factors of the Management of Protected Areas using SWOT Analysis, *Journal of Ecology*, Volume 7, No. 1, Pages 55–68. https://doi.org/10.4236/oje.2017.71005

Salomone, R., (2008), Integrated Management Systems: Experiences in Italian Organizations, *Journal of Cleaner Production*, Volume 16, No. 16, Pages 1786–1806. https://doi.org/10.1016/j.jclepro.2007.12.003

Sima, H., Jana, P., Panghal, D., (2019), Feasibility of Using Simulation Technique for Line Balancing in Apparel Industry, *Procedia Manufacturing*, Volume 30, Pages 300–307. https://doi.org/10.1016/j.promfg.2019.02.043

Tulcidas, A., Nascimento, S., Santos, B., Alvarez, C., Pawlowski, S., Rocha, F., (2019), Statistical Methodology for Scale-up of an Anti-solvent Crystallization Process in the Pharmaceutical Industry, *Separation and Purification Technology*, Volume 213, Pages 56–62. https://doi.org/10.1016/j.seppur.2018.12.019

Zhou, X., (2016), Mechanism Design Theory: The Development in Economics and Management, *Journal of Business and Management*, Volume 4, No. 2, Pages 345–348. https://doi.org/10.4236/ojbm.2016.42036

Appendix: Assessment Questionnaire for Generalizing the Proposed Model

Respondent's personal and organizational characteristics				
Name & Surname (optional)				
Education Level	Bachelor's ☐	Master's ☐	Professional ☐	
Organizational Position	Manager ☐		Expert ☐	
Experience in the field of industry (Year)	Less than 10 ☐	Between 10 to 20 ☐	More than 20 ☐	
Level of familiarity with Quality Engineering Techniques (QET)	Elementary ☐			
	Intermediate ☐			
	Advanced ☐			
Level of familiarity with Processes in the industrial organizations	Very High ☐	High ☐	Moderate ☐	Low ☐	Very Low ☐
The volume of product manufacturing	Mass Production ☐	Semi Mass Production ☐	Prototype production ☐	Individual ☐	other ☐

1. How comprehensive is the proposed model?								
1	2	3	4	5	6	7	8	9
2. How feasible is the proposed model?								
1	2	3	4	5	6	7	8	9
3. How flexible is the proposed model?								
1	2	3	4	5	6	7	8	9
4. How stable is the proposed model?								
1	2	3	4	5	6	7	8	9
5. How powerful is the proposed model?								
1	2	3	4	5	6	7	8	9

Guidelines for the chapters' exercises

CHAPTER 1:

1.1. The main specifications of QET consist of:

- Ease of learning by audiences
- Comprehensive interpretation
- Strong graphical presentation
- Estimation in change point
- Attractions to stakeholders
- Information interchange
- Technique consistency
- Mathematical analysis
- Power of assessment
- Technique flexibility
- Technique validity
- Power of upgrade

N	Specifications of Technique	Productivity	Sustainability
1	Ease of learning by audiences		
2	Comprehensive interpretation		
3	Strong graphical presentation		
4	Estimation in change point		
5	Attractions to stakeholders		
6	Information interchange		
7	Technique consistency		
8	Mathematical analysis		
9	Power of assessment		
10	Technique flexibility		
11	Technique validity		
12	Power of upgrade		

Regard to productivity and sustainability concepts explained in Chapter 1, complete the above table.

1.2. The main components of SCM consist of:

- Customers = Determining what customers want
- Forecasting = Predicting quantity and timing of demand
- Designing = Time and specifications that customers want
- Processing = Controlling quality and scheduling work
- Inventory = Meeting demand while managing inventory costs
- Purchasing = Evaluating suppliers and supporting operations
- Suppliers = Monitoring suppliers quality, delivery and relations
- Location = Determining location of all related facilities
- Logistics = Deciding how to best move and store materials

N	Main components of SCM		Productivity	Sustainability
1	Customers	Determining what customers want		
2	Forecasting	Predicting the quantity and timing of demand		
3	Designing	Time and specifications that customers want		
4	Processing	Controlling quality and scheduling work		
5	Inventory	Meeting demand while managing inventory costs		
6	Purchasing	Evaluating suppliers and supporting operations		
7	Suppliers	Monitoring suppliers quality, delivery, and relations		
8	Location	Determining the location of all related facilities		
9	Logistics	Deciding how to best move and store materials		

Regard to productivity and sustainability concepts explained in Chapter 1, complete the above table.

CHAPTER 2:

2.1. Reliability and validity are concepts used to evaluate the quality of research. They indicate how well a method, technique or test measures something. Reliability is about the consistency of a measure, and validity is about the accuracy of a measure. It's important to consider reliability and validity when you are creating your research design, planning your methods, and writing up your results, especially in quantitative research.

Reliability refers to how consistently a method measures something. If the same result can be consistently achieved by using the same methods under the same circumstances, the measurement is considered reliable.

Validity refers to how accurately a method measures what it is intended to measure. If research has high validity that means it produces results that correspond to real properties, characteristics, and variations in the physical or social world. High reliability is one indicator that measurement is valid. If a method is not reliable, it probably isn't valid.

CHAPTER 3:

3.1. Many researchers introduce the main components of Supply Chain Management (SCM) in their reference books as follows:

- Customers = Determining what customers want
- Forecasting = Predicting quantity and timing of demand

- Designing = Time and specifications that customers want
- Processing = Controlling quality and scheduling work
- Inventory = Meeting demand while managing inventory costs
- Purchasing = Evaluating suppliers and supporting operations
- Suppliers = Monitoring suppliers quality, delivery and relations
- Location = Determining location of all related facilities
- Logistics = Deciding how to best move and store materials

By forming two assessment matrices between the main components of SCM and QET, the strongest Quality Engineering Techniques including statistical and non-statistical can be defined and selected.

There are two matrices in this project:

- Matrix A is presented for statistical techniques.
- Matrix B is presented for non-statistical techniques.

For completing these matrices, it is required to the number of sufficient respondents (sample size), calculating the average of data in each house of matrices, and estimating the total average number for each row and column. Research reliability must not be neglected.

After selecting the strongest Quality Engineering Techniques including statistical or non-statistical, the final model will be considered.

Matrix A

Main Components of Supply Chain Management (SCM)	Statistical Techniques based on ISO10017	Descriptive Statistics	Design of Experiments	Statistical Hypothesis Tests	Process Capability Analysis	Regression Analysis	Reliability Analysis	Sampling	Simulation	Statistical Process Control	Statistical Tolerances	Time Series Analysis
Customers												
Forecasting												
Designing												
Processing												
Inventory												
Purchasing												
Suppliers												
Location												
Logistics												
Total Average												

Matrix B

Main Components of Supply Chain Management (SCM)	Non-Statistical Techniques	Quality Function Deployment	Value Engineering	Value Stream Mapping	Work Flow Analysis	Cost of Quality	Failure Mode Effects Analysis	Designing Failure Mode Effects Analysis	Production Failure Mode Effects Analysis	Voice of the Customer	Brain Storming	Nominal Group
Customers												
Forecasting												
Designing												
Processing												
Inventory												
Purchasing												
Suppliers												
Location												
Logistics												
Total Average												

3.2. Many researchers introduce the main domains of Industrial Engineering (IE) in their reference books as follows (Karimi Gavareshki et al., 2018):

- Manufacturing Resources Management
- Manufacturing Processes Design
- Planning and Manufacturing Control
- Manufacturing Timing Management
- Manufacturing Risk Management
- Product Reliability Management
- Production Waste Management
- Quality System Management
- Projects Control Management
- Equipment Locating Management
- Materials Transportation Management
- Productions Transportation Management

By forming two assessment matrices between the main domains of IE and QET, the strongest Quality Engineering Techniques including statistical and non-statistical can be defined and selected.

There are two matrices in this project:

- Matrix A is presented for statistical techniques.
- Matrix B is presented for non-statistical techniques.

For completing these matrices, it is required to the number of sufficient respondents (sample size), calculating the average of data in each house of matrices, and estimating the total average number for each row and column. Research reliability must not be neglected.

After selecting the strongest Quality Engineering Techniques including statistical or non-statistical, the final model will be considered.

Matrix A

Main Domains of Industrial Engineering (IE)	Statistical Techniques based on ISO10017	Descriptive Statistics	Design of Experiments	Statistical Hypothesis Tests	Process Capability Analysis	Regression Analysis	Reliability Analysis	Sampling	Simulation	Statistical Process Control	Statistical Tolerances	Time Series Analysis
Manufacturing Resources Management												
Manufacturing Processes Design												
Planning and Manufacturing Control												
Manufacturing Timing Management												
Manufacturing Risk Management												
Product Reliability Management												
Production Waste Management												
Quality System Management												
Projects Control Management												
Equipment Locating Management												
Materials Transportation Management												
Productions Transportation Management												
Total Average												

Matrix B

Main Domains of Industrial Engineering (IE)	Non-Statistical Techniques	Quality Function Deployment	Value Engineering	Value Stream Mapping	Work Flow Analysis	Cost of Quality	Failure Mode Effects Analysis	Designing Failure Mode Effects Analysis	Production Failure Mode Effects Analysis	Voice of the Customer	Brain Storming	Nominal Group
Manufacturing Resources Management												
Manufacturing Processes Design												
Planning and Manufacturing Control												
Manufacturing Timing Management												
Manufacturing Risk Management												
Product Reliability Management												
Production Waste Management												
Quality System Management												
Projects Control Management												
Equipment Locating Management												
Materials Transportation Management												
Productions Transportation Management												
Total Average												

CHAPTER 4:

4.1. Section (A) belongs to the analysis of the results in order of importance and section (B) belongs to the analysis of the results in performance.

(A): The mean and median of the importance section in the below figure are shown:

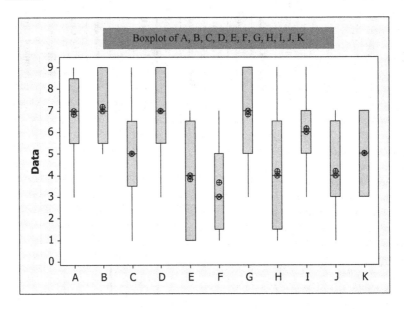

As can be seen, the mean and median of the data related to the importance of statistical techniques of quality engineering implemented in QMS have acceptable values above 50% and equals (5.44). Moreover, the overlap of the mean and median of these data are high. It is clear the average of data for the design and analysis of experiments is the highest, and for the reliability analysis is the lowest. The data variance for the simulation technique is the highest, and for the design and analysis of experiments is the lowest. This means that respondents in the simulation technique have the lowest agreement, which can be due to various reasons, including a lack of understanding of the application of this technique. On the contrary, for the design and analysis of experiments, which the highest average of data was obtained for it, the lowest variance has been obtained, which can be due to the more recognition of respondents to this technique, and, as a result, the greatest agreement has been formed on it.

(B): The mean and median of the performance section in the below figure are shown:

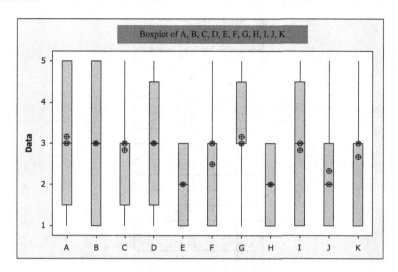

The overlap of the mean and median of these data are high. Thus, it is clear from the shape of the figure that the current state of the performance of these techniques in the sample industry is lower than the average. The average of data for the descriptive statistics, and sampling techniques are the highest, and for the regression, and simulation techniques are the lowest. As a result, in terms of respondents, the descriptive statistics, and sampling techniques have the highest practical applications, and the regression analysis and simulation techniques have the lowest practical and current application in this sample industry. The highest variance is for the design and analysis of experiments, and the lowest variance is for the regression, and simulation techniques. Unlike the similar situation with regard to the importance of these techniques for practical application (existing situation), respondents have the lowest agreement on the design and analysis of experiments (while the highest agreement was about their importance), and in the case of the regression and simulation techniques have the highest agreement on their practical application (existing situation), while the lowest agreement was about the importance of simulation techniques.

Additional explanations and implemented mathematical calculations to find the best Statistical Distribution Function (SDF)

N	Failure Time (hour)
1	1.25
2	2.35
3	3.75
4	5.50
5	4.55
6	7.85
7	6.50
8	8.85
9	6.45
10	7.50
11	10.25
12	12.65
13	14.35
14	16.25
15	18.00
16	19.95
17	21.00
18	22.85
19	23.95
20	24.15
21	32.75
22	36.05
23	51.12
24	77.22
25	120.00
26	138.50
27	155.55
28	177.75
29	210.25
30	228.00

Exponential Distribution Function				Chi-square(χ^2)
$f(t,\Lambda) = \Lambda e^{-\Lambda t}$		Λ parameter	0.0197	27.8646
P(1)	0.0076	P(0)	P(24.99)	P(1)=P(0)-P(24.99)	
FP$_1$	0.6667	0.0197	0.0120	0.0076	
(P(1)-FP$_1$)^2	0.4343	(P(1)-FP$_1$)^2/FP$_1$		0.6515	
P(2)	0.0047	P(25)	P(49.99)	P(2)=P(25)-P(49.99)	
FP$_2$	0.0667	0.0120	0.0074	0.0047	
(P(2)-FP$_2$)^2	0.0038	(P(2)-FP$_2$)^2/FP$_2$		0.0577	
P(3)	0.0029	P(50)	P(74.99)	P(3)=P(50)-P(74.99)	
FP$_3$	0.0333	0.0074	0.0045	0.0029	
(P(3)-FP$_3$)^2	0.0009	(P(3)-FP$_3$)^2/FP$_3$		0.0279	
P(4)	0.0017	P(75)	P(99.99)	P(4)=P(75)-P(99.99)	
FP$_4$	0.0333	0.0045	0.0028	0.0017	
(P(4)-FP$_4$)^2	0.0010	(P(4)-FP$_4$)^2/FP$_4$		0.0299	
P(5)	0.0011	P(100)	P(124.99)	P(5)=P(100)-P(124.99)	
FP$_5$	0.0333	0.0028	0.0017	0.0011	
(P(5)-FP$_5$)^2	0.0010	(P(5)-FP$_5$)^2/FP$_5$		0.0312	
P(6)	0.0005	P(125)	P(144.99)	P(6)=P(125)-P(144.99)	
FP$_6$	0.0333	0.0017	0.0011	0.0005	
(P(6)-FP$_6$)^2	0.0011	(P(6)-FP$_6$)^2/FP$_6$		0.0322	
P(7)	0.0004	P(150)	P(174.99)	P(7)=P(150)-P(174.99)	
FP$_7$	0.0333	0.0010	0.0006	0.0004	
(P(7)-FP$_7$)^2	0.0011	(P(7)-FP$_7$)^2/FP$_7$		0.0325	
P(8)	0.0002	P(175)	P199.99)	P(8)=P(175)-P(199.99)	
FP$_8$	0.0333	0.0006	0.0004	0.0002	
(P(8)-FP$_8$)^2	0.0011	(P(8)-FP$_8$)^2/FP$_8$		0.0328	
P(9)	0.0001	P(200)	P(224.99)	P(9)=P(200)-P(224.99)	
FP$_9$	0.0333	0.0004	0.0002	0.0001	
(P(9)-FP$_9$)^2	0.0011	(P(9)-FP$_9$)^2/FP$_9$		0.0330	
P(10)	0.0000	P(225)	P(249.99)	P(10)=P(225)-P(249.99)	
FP$_{10}$	0.0000	0.0002	0.0001	0.0001	
(P(10)-FP$_{10}$)^2	0.0000	(P(10)-FP$_{10}$)^2/FP$_{10}$		0.0000	

$\chi^2 = 30 \times \sum (P(i)-FP_i)^2/FPi$

27.8646

Ultra Exponential Distribution Function		k parameter		0.25		Chi-square($χ^2$)
$f(t,Λ,k) = 2Λk^2e^{-2Λkt} +$ $2Λ(1-k)^2e^{-2Λ(1-k)t}$		Λ parameter		0.0197		28.5759
		2Λk	0.0098	2Λ(1-k)	0.0295	
P(1)	0.0121	P(0)	P(24.99)	P(0)-P(24.99)		28.5759
FP_1	0.6667	0.0246	0.0125	0.0121		
$(P(1)-FP_1)^2$	0.4285	$(P(1)-FP_1)^2/FP_1$		0.6427		
P(2)	0.0059	P(25)	P(49.99)	P(25)-P(49.99)		
FP_2	0.0667	0.0125	0.0066	0.0059		
$(P(2)-FP_2)^2$	0.0037	$(P(2)-FP_2)^2/FP_2$		0.0553		
P(3)	0.0030	P(50)	P(74.99)	P(50)-P(74.99)		
FP_3	0.0333	0.0066	0.0036	0.0030		
$(P(3)-FP_3)^2$	0.0009	$(P(3)-FP_3)^2/FP_3$		0.0277		
P(4)	0.0015	P(75)	P(99.99)	P(75)-P(99.99)		
FP_4	0.0333	0.0036	0.0021	0.0015		
$(P(4)-FP_4)^2$	0.0010	$(P(4)-FP_4)^2/FP_4$		0.0304		
P(5)	0.0008	P(100)	P(124.99)	P(100)-P(124.99)		
FP_5	0.0333	0.0021	0.0013	0.0008		
$(P(5)-FP_5)^2$	0.0011	$(P(5)-FP_5)^2/FP_5$		0.0317		
P(6)	0.0004	P(125)	P(144.99)	P(125)-P(144.99)		$χ^2 = 30 \times \sum (P(i)-FP_i)^2/FP_i$
FP_6	0.0333	0.0013	0.0009	0.0004		
$(P(6)-FP_6)^2$	0.0011	$(P(6)-FP_6)^2/FP_6$		0.0326		
P(7)	0.0003	P(150)	P(174.99)	P(150)-P(174.99)		
FP_7	0.0333	0.0008	0.0006	0.0003		
$(P(7)-FP_7)^2$	0.0011	$(P(7)-FP_7)^2/FP_7$		0.0328		
P(8)	0.0002	P(175)	P199.99)	P(175)-P(199.99)		
FP_8	0.0333	0.0006	0.0004	0.0002		
$(P(8)-FP_8)^2$	0.0011	$(P(8)-FP_8)^2/FP_8$		0.0330		
P(9)	0.0001	P(200)	P(224.99)	P(200)-P(224.99)		
FP_9	0.0333	0.0004	0.0003	0.0001		
$(P(9)-FP_9)^2$	0.0011	$(P(9)-FP_9)^2/FP_9$		0.0331		
P(10)	0.0001	P(225)	P(249.99)	P(225)-P(249.99)		
FP_{10}	0.0333	0.0003	0.0002	0.0001		
$(P(10)-FP_{10})^2$	0.0011	$(P(10)-FP_{10})^2/FP_{10}$		0.0332		

Gamma Distribution Function			β parameter	100	Chi-square(χ^2)
$f(t,\alpha,\beta) = [\beta^\alpha\Gamma(\alpha)]^{-1}t^{\alpha-1}$ exp-(t/β)			α parameter	2	30.1277
$P(1)$	0.0019	$P(0)$	$P(24.99)$	$P(0)$-$P(24.99)$	30.1277
FP_1	0.6667	0.0000	0.0019	0.0019	
$(P(1)-FP_1)^2$	0.4470	$(P(1)-FP_1)^2/FP_1$		0.6706	
$P(2)$	0.0011	$P(25)$	$P(49.99)$	$P(25)$-$P(49.99)$	
FP_2	0.0667	0.0019	0.0030	0.0011	
$(P(2)-FP_2)^2$	0.0046	$(P(2)-FP_2)^2/FP_2$		0.0689	
$P(3)$	0.0005	$P(50)$	$P(74.99)$	$P(50)$-$P(74.99)$	
FP_3	0.0333	0.0030	0.0035	0.0005	
$(P(3)-FP_3)^2$	0.0011	$(P(3)-FP_3)^2/FP_3$		0.0344	
$P(4)$	0.0001	$P(75)$	$P(99.99)$	$P(75)$-$P(99.99)$	
FP_4	0.0333	0.0035	0.0037	0.0001	
$(P(4)-FP_4)^2$	0.0011	$(P(4)-FP_4)^2/FP_4$		0.0336	
$P(5)$	0.0001	$P(100)$	$P(124.99)$	$P(100)$-$P(124.99)$	
FP_5	0.0333	0.0037	0.0036	0.0001	
$(P(5)-FP_5)^2$	0.0011	$(P(5)-FP_5)^2/FP_5$		0.0331	
$P(6)$	0.0002	$P(125)$	$P(144.99)$	$P(125)$-$P(144.99)$	
FP_6	0.0333	0.0036	0.0034	0.0002	
$(P(6)-FP_6)^2$	0.0011	$(P(6)-FP_6)^2/FP_6$		0.0330	
$P(7)$	0.0003	$P(150)$	$P(174.99)$	$P(150)$-$P(174.99)$	
FP_7	0.0333	0.0033	0.0030	0.0003	
$(P(7)-FP_7)^2$	0.0011	$(P(7)-FP_7)^2/FP_7$		0.0327	
$P(8)$	0.0003	$P(175)$	$P199.99)$	$P(175)$-$P(199.99)$	
FP_8	0.0333	0.0030	0.0027	0.0003	
$(P(8)-FP_8)^2$	0.0011	$(P(8)-FP_8)^2/FP_8$		0.0327	
$P(9)$	0.0003	$P(200)$	$P(224.99)$	$P(200)$-$P(224.99)$	
FP_9	0.0333	0.0027	0.0024	0.0003	
$(P(9)-FP_9)^2$	0.0011	$(P(9)-FP_9)^2/FP_9$		0.0327	
$P(10)$	0.0003	$P(225)$	$P(249.99)$	$P(225)$-$P(249.99)$	
FP_{10}	0.0333	0.0024	0.0021	0.0003	
$(P(10)-FP_{10})^2$	0.0011	$(P(10)-FP_{10})^2/FP_{10}$		0.0327	

$\chi^2=30 \times \sum (P(i)-FP_i)^2/FPi$

Weibull Distribution Function	β parameter			1000		Chi-square(χ^2)
$f(t,\alpha,\beta) = \alpha\beta^{-\alpha}t^{\alpha-1}\exp-(t/\beta)^\alpha$	α parameter			4		30.0035
$P(1)$	0.00000006	$P(0)$	$P(24.99)$	$P(0)-P(24.99)$		30.0035
FP_1	0.6667	0.0000	0.0000	0.00000006		
$(P(1)-FP_1)^2$	0.4444	$(P(1)-FP_1)^2/FP_1$		0.6667		
$P(2)$	0.00000044	$P(25)$	$P(49.99)$	$P(25)-P(49.99)$		
FP_2	0.0667	0.0000	0.0000	0.00000044		
$(P(2)-FP_2)^2$	0.0044	$(P(2)-FP_2)^2/FP_2$		0.0667		
$P(3)$	0.00000119	$P(50)$	$P(74.99)$	$P(50)-P(74.99)$		
FP_3	0.0333	0.0000	0.0000	0.00000119		
$(P(3)-FP_3)^2$	0.0011	$(P(3)-FP_3)^2/FP_3$		0.0333		
$P(4)$	0.00000036	$P(75)$	$P(99.99)$	$P(75)-P(99.99)$		
FP_4	0.0333	0.0000	0.0000	0.00000036		
$(P(4)-FP_4)^2$	0.0011	$(P(4)-FP_4)^2/FP_4$		0.0333		
$P(5)$	0.00000381	$P(100)$	$P(124.99)$	$P(100)-P(124.99)$		
FP_5	0.0333	0.0000	0.0000	0.00000381		
$(P(5)-FP_5)^2$	0.0011	$(P(5)-FP_5)^2/FP_5$		0.0333	$\chi^2=30 \times \sum (P(i)-FP_i)^2/FPi$	
$P(6)$	0.00000438	$P(125)$	$P(144.99)$	$P(125)-P(144.99)$		
FP_6	0.0333	0.0000	0.0000	0.00000438		
$(P(6)-FP_6)^2$	0.0011	$(P(6)-FP_6)^2/FP_6$		0.0333		
$P(7)$	0.00000792	$P(150)$	$P(174.99)$	$P(150)-P(174.99)$		
FP_7	0.0333	0.0000	0.0000	0.00000792		
$(P(7)-FP_7)^2$	0.0011	$(P(7)-FP_7)^2/FP_7$		0.0333		
$P(8)$	0.00001053	$P(175)$	$P199.99)$	$P(175)-P(199.99)$		
FP_8	0.0333	0.0000	0.0000	0.00001053		
$(P(8)-FP_8)^2$	0.0011	$(P(8)-FP_8)^2/FP_8$		0.0334		
$P(9)$	0.00001349	$P(200)$	$P(224.99)$	$P(200)-P(224.99)$		
FP_9	0.0333	0.0000	0.0000	0.00001349		
$(P(9)-FP_9)^2$	0.0011	$(P(9)-FP_9)^2/FP_9$		0.0334		
$P(10)$	0.00001680	$P(225)$	$P(249.99)$	$P(225)-P(249.99)$		
FP_{10}	0.0333	0.0000	0.0001	0.00001680		
$(P(10)-FP_{10})^2$	0.0011	$(P(10)-FP_{10})^2/FP_{10}$		0.0334		

Normal Distribution Function		σ parameter		66.6657	Chi-square(χ^2)
$f(t,\mu,\sigma)=(2\pi)^{-1/2}\sigma^{-1}\exp-[(t-\mu)^2/2\sigma^2]$		m parameter		50.8333	29.7544
P(1)	0.0011	P(0)	P(24.99)	P(0)-P(24.99)	29.7544
FP_1	0.6667	0.0045	0.0056	0.0011	
$(P(1)-FP_1)^2$	0.4459	$(P(1)-FP_1)^2/FP_1$		0.6688	
P(2)	0.0004	P(25)	P(49.99)	P(25)-P(49.99)	
FP_2	0.0667	0.0056	0.0060	0.0004	
$(P(2)-FP_2)^2$	0.0045	$(P(2)-FP_2)^2/FP_2$		0.0675	
P(3)	0.0004	P(50)	P(74.99)	P(50)-P(74.99)	
FP_3	0.0333	0.0060	0.0056	0.0004	
$(P(3)-FP_3)^2$	0.0011	$(P(3)-FP_3)^2/FP_3$		0.0326	
P(4)	0.0010	P(75)	P(99.99)	P(75)-P(99.99)	
FP_4	0.0333	0.0056	0.0046	0.0010	
$(P(4)-FP_4)^2$	0.0010	$(P(4)-FP_4)^2/FP_4$		0.0313	
P(5)	0.0013	P(100)	P(124.99)	P(100)-P(124.99)	
FP_5	0.0333	0.0046	0.0032	0.0013	
$(P(5)-FP_5)^2$	0.0010	$(P(5)-FP_5)^2/FP_5$		0.0307	
P(6)	0.0010	P(125)	P(144.99)	P(125)-P(144.99)	
FP_6	0.0333	0.0032	0.0022	0.0010	
$(P(6)-FP_6)^2$	0.0010	$(P(6)-FP_6)^2/FP_6$		0.0313	
P(7)	0.0009	P(150)	P(174.99)	P(150)-P(174.99)	
FP_7	0.0333	0.0020	0.0011	0.0009	
$(P(7)-FP_7)^2$	0.0011	$(P(7)-FP_7)^2/FP_7$		0.0315	
P(8)	0.0006	P(175)	P199.99)	P(175)-P(199.99)	
FP_8	0.0333	0.0011	0.0005	0.0006	
$(P(8)-FP_8)^2$	0.0011	$(P(8)-FP_8)^2/FP_8$		0.0322	
P(9)	0.0003	P(200)	P(224.99)	P(200)-P(224.99)	
FP_9	0.0333	0.0005	0.0002	0.0003	
$(P(9)-FP_9)^2$	0.0011	$(P(9)-FP_9)^2/FP_9$		0.0328	
P(10)	0.0001	P(225)	P(249.99)	P(225)-P(249.99)	
FP_{10}	0.0333	0.0002	0.0001	0.0001	
$(P(10)-FP_{10})^2$	0.0011	$(P(10)-FP_{10})^2/FP_{10}$		0.0331	

$\chi^2 = 30 \times \sum (P(i)-FP_i)^2/FPi$

Log-normal Distribution Function		σ parameter		1.3842	Chi-square(χ^2)
$f(t,\mu,\sigma)=0.4343(2\pi)^{-1/2}(t\sigma)^{-1}$ $\exp-[(\log(t)-\mu)^2/2\sigma^2]$		m parameter		3.0141	28.4382
P(1)	0.0155	P(1)	P(24.99)	P(0)-P(24.99)	28.4382
		0.0000	3.2185		
FP_1	0.6667	0.0269	0.0114	0.0155	
$(P(1)-FP_1)^2$	0.4240	$(P(1)-FP_1)^2/FP_1$		0.6360	
P(2)	0.0067	P(25)	P(49.99)	P(25)-P(49.99)	
		3.2189	3.9118		
FP_2	0.0667	0.0114	0.0047	0.0067	
$(P(2)-FP_2)^2$	0.0036	$(P(2)-FP_2)^2/FP_2$		0.0539	
P(3)	0.0022	P(50)	P(74.99)	P(50)-P(74.99)	
		3.9120	4.3174		
FP_3	0.0333	0.0047	0.0025	0.0022	
$(P(3)-FP_3)^2$	0.0010	$(P(3)-FP_3)^2/FP_3$		0.0291	
P(4)	0.0010	P(75)	P(99.99)	P(75)-P(99.99)	
		4.3175	4.6051		
FP_4	0.0333	0.0025	0.0015	0.0010	
$(P(4)-FP_4)^2$	0.0010	$(P(4)-FP_4)^2/FP_4$		0.0314	
P(5)	0.0005	P(100)	P(124.99)	P(100)-P(124.99)	
		4.6052	4.8282		
FP_5	0.0333	0.0015	0.0010	0.0005	
$(P(5)-FP_5)^2$	0.0011	$(P(5)-FP_5)^2/FP_5$		0.0323	
P(6)	0.0002	P(125)	P(144.99)	P(125)-P(144.99)	
		4.8283	4.9767		
FP_6	0.0333	0.0010	0.0007	0.0002	
$(P(6)-FP_6)^2$	0.0011	$(P(6)-FP_6)^2/FP_6$		0.0328	$\chi^2=30 \times \sum (P(i)-FP_i)^2/FPi$
P(7)	0.0002	P(150)	P(174.99)	P(150)-P(174.99)	
		5.0106	5.1647		
FP_7	0.0333	0.0007	0.0005	0.0002	
$(P(7)-FP_7)^2$	0.0011	$(P(7)-FP_7)^2/FP_7$		0.0330	
P(8)	0.0001	P(175)	P199.99)	P(175)-P(199.99)	
		5.1648	5.2983		
FP_8	0.0333	0.0005	0.0004	0.0001	
$(P(8)-FP_8)^2$	0.0011	$(P(8)-FP_8)^2/FP_8$		0.0331	
P(9)	0.0001	P(200)	P(224.99)	P(200)-P(224.99)	
		5.2983	5.4161		
FP_9	0.0333	0.0004	0.0003	0.0001	
$(P(9)-FP_9)^2$	0.0011	$(P(9)-FP_9)^2/FP_9$		0.0332	
P(10)	0.0001	P(225)	P(249.99)	P(225)-P(249.99)	
		5.4161	5.5214		
FP_{10}	0.0333	0.0003	0.0002	0.0001	
$(P(10)-FP_{10})^2$	0.0011	$(P(10)-FP_{10})^2/FP_{10}$		0.0332	

Type of Statistical Distribution Function	Formula	χ^2
Exponential Distribution Function	$f(t,\lambda)=\lambda e^{-\lambda t}$	27.8646
Ultra Exponential Distribution Function	$f(t,\lambda,k)=2\lambda k^2 e^{-2\lambda kt}+2\lambda(1-k)^2 e^{-2\lambda(1-k)t}$	28.5759
Gamma Function	$f(t,\alpha,\beta)=\dfrac{1}{\beta^{\alpha}\Gamma(\alpha)}\,t^{\alpha-1}e^{-\frac{t}{\beta}}$	30.1277
Weibull Function	$f(t,\alpha,\beta)=\dfrac{\alpha}{\beta^{\alpha}}\,t^{\alpha-1}e^{-(\frac{t}{\beta})\alpha}$	30.0035
Normal Function	$f(t,\mu,\sigma)=\dfrac{1}{\sigma\sqrt{2\pi}}\,e^{-[\frac{(t-\mu)^2}{2\sigma^2}]}$	29.7544
Log-normal Function	$f(t,\mu,\sigma)=\dfrac{0.4343}{t\sigma\sqrt{2\pi}}\,e^{-[\frac{(\log(t)-\mu)^2}{2\sigma^2}]}$	28.4382
Maximum Chi-Square Permissible	$\chi^2 0.001,18$	27.8680

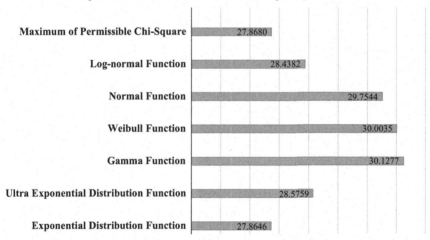

Chi-square test values of the statistical distribution functions in compared with maximum of two-tailed chi-square permissible

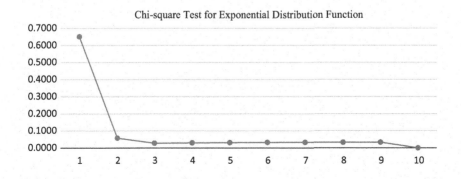

Chi-square Test for Exponential Distribution Function

Index

Taylor & Francis eBooks

www.taylorfrancis.com

A single destination for eBooks from Taylor & Francis
with increased functionality and an improved user
experience to meet the needs of our customers.

90,000+ eBooks of award-winning academic content in
Humanities, Social Science, Science, Technology, Engineering,
and Medical written by a global network of editors and authors.

TAYLOR & FRANCIS EBOOKS OFFERS:

A streamlined
experience for
our library
customers

A single point
of discovery
for all of our
eBook content

Improved
search and
discovery of
content at both
book and
chapter level

REQUEST A FREE TRIAL
support@taylorfrancis.com

 Routledge
Taylor & Francis Group

 CRC Press
Taylor & Francis Group

Printed in the United ...
... published, Taylor Francis or us ... services.

Printed in the United States
by Baker & Taylor Publisher Services